THE HANDY Life in the OCEAN GK BOOK

THE HANDY Life in the OCEAN GK BOOK

Thomas E. Svarney & Patricia Barnes-Svarney

JAICO PUBLISHING HOUSE

Mumbai • Delhi • Bangalore • Kolkata
Hyderabad • Chennai • Ahmedabad • Bhopal

Published by Jaico Publishing House
121 Mahatma Gandhi Road
Mumbai - 400 001
jaicopub@vsnl.com
www.jaicobooks.com

Published in arrangement with
Visible Ink Press®
42015 Ford Rd., #208
Canton, MI 48187-3669

THE HANDY LIFE IN THE OCEAN GK BOOK
ISBN 81-7992-464-5

First Jaico Impression: 2007

Printed by
Printed by Saurabh Printers Pvt. Ltd., Noida

Contents

Introduction

It is difficult for those of us who live inland to understand the immensity and effects of the world's oceans. These vast bodies of water take up 70 percent of the surface area of our globe, with a volume close to 350 million cubic miles. And at present, ours is the only known planet in the solar system to contain such waters.

Even those of us living in the middle of a continent are affected by these immense bodies of water in a direct or indirect way. For example, the Sun heats the surface of the oceans, evaporating waters to form clouds and mixing the air currents to produce our weather systems. The Sun's heat also stirs the surface waters, creating waves and currents that sculpt our shorelines. Seasonal temperature changes of the ocean waters alternately increase or decrease populations of marine organisms, including many that humans harvest for food.

Most importantly, without the oceans, there would be no life. It was in the oceans, billions of years ago, that life began probably in shallower waters or around deep hydrothermal vents. Either way, life evolved under water, eventually reaching the edges of the oceans and making its way to land. As amazing as it seems, our entire population of terrestrial and marine organisms all had a common beginning in our oceans.

The Handy Life In the Ocean GK Book addresses present answers to the most common questions about our oceans covering features and organisms from the shoreline to the open ocean. Here, we examine the physical attributes of the oceans, marine animals and plants, and finally, the human ties to (and our effect on) the oceans.

Many people call the oceans our most important natural resource. The waters have furnished humans with food for centuries from a plethora of fish species to certain marine plants. But people have also adversely affected the oceans: Over-fishing, coastal erosion caused by development, and pollution threaten this natural habitat. We need to keep the oceans and their bounty if we are to survive as a species.

The world ocean continues to hold many secrets; there are numerous questions still to be answered. For example, what kinds of organisms are found in the ocean's deepest waters? How many fish thought to be extinct are actually still alive? What species are important to coral reef growth? How do microorganisms live in the coldest waters of the Arctic? Other questions have to do with the connections and interdependencies between humans and the ocean. For example, are plankton (one of the most important organisms in the marine food chain) able to withstand environmental stresses such as the ozone hole? Can humans continue to harvest the oceans without disrupting the balance between the organisms and their environment? Scientists hope to answer some of these questions in the near future, not only by using better technology that allows humans to dive into and explore the oceans for longer periods, but through new satellites that watch the global waters tracking changes over time.

We hope these pages describe an underwater world that will inform and inspire you, and, perhaps, engage your interest enough that you'll want to discover even more about the wonders of the ocean. This is the mysterious and largely unexplored territory where life originated and it will be an integral part of our future.

THE ANCIENT OCEANS

THE GEOLOGIC TIME SCALE

What is the **geologic time scale**?

The geologic time scale is a tool scientists developed to represent the Earth's history—from the formation of the planet, about 4.55 billion years ago, to the present. The major divisions on the scale are listed after the Cambrian period (at the beginning of the Paleozoic era). The reason is that major fossils do not show up in the Earth's rock layers until after what is called the great Cambrian Explosion, when life began to flourish in the oceans.

What does the **geologic time scale** look like?

The following represents one version of the geologic time scale (there are other scales with different dates and names, usually depending on the country).

Eon	Era	Period	Epoch
Phanerozoic eon (began 544 million years ago)	Cenozoic era	Quaternary period (1.8 million years ago to the present)	Holocene epoch—11,000 years ago to present
			Pleistocene (glacial) epoch—1.8 million years to 11,000 years ago

Eon	Era	Period	Epoch
Phanerozoic eon, (continued)	(Cenozoic era, continued)	Tertiary period (65 to 1.8 million years ago)	Pliocene epoch—5 to 1.8 million years ago
			Miocene epoch—23 to 5 million years ago
			Oligocene epoch—38 to 23 million years ago
			Eocene epoch—54 to 38 million years ago
			Paleocene epoch—65 to 54 million years ago
	Mesozoic era	Cretaceous period (146 to 65 million years ago)	
		Jurassic period (208 to 146 million years ago)	
		Triassic period (245 to 208 million years ago)	
	Paleozoic era	Permian period (286 to 245 million years ago)	
		Carboniferous period (360 to 286 million years ago; it can be broken up into the Pennsylvanian period, 325 to 286 million years ago; and Mississippian period, 360 to 325 million years ago)	
		Devonian period (410 to 360 million years ago)	

Eon	Era	Period	Epoch
Phanerozoic eon, (continued)	(Paleozoic era, continued)	Silurian period (440 to 410 million years ago)	
		Ordovician period (505 to 440 million years ago)	
		Cambrian period (544 to 505 million years ago)	
Precambrian eon (4.55 billion years to 544 million years ago)			

Who first **subdivided Earth's** long **history**?

William Smith (1769–1839), an English canal engineer, was one of the first to subdivide the Earth's long history. His 1815 geological map of England and Wales established a practical system of stratigraphy—the geology of the layers of the Earth's crust; stratigraphy is the basis of the geologic time scale. Smith showed that certain layers of Mesozoic rock in England could be identified by their specific fossils.

An international geologic time scale was drawn up between 1820 and 1870; the era divisions labeled Paleozoic ("ancient life"), Mesozoic ("middle life"), and Cenozoic ("recent life") were established around 1840. By the end of the nineteenth century, the eras were further subdivided into periods, epochs, zones (now usually called ages), and subzones (usually called subages). And by the mid-twentieth century, divisions were more precise, as scientists began using radiometric dating techniques to determine the absolute ages of the rock layers.

How can a person **visualize** the extent of the **geologic time scale**?

The geologic time scale represents billions of years—a time span that is almost beyond human comprehension. One of the best ways to get a handle on how much time the scale actually represents was suggested

The fossil of an early plant, dating from the Devonian period, or between 360 and 410 million years old. *CORBIS/James L. Amos*

by author John McPhee, in his book *Basin and Range*: Stand with your arms held straight out to each side. The extent of the Earth's history, as represented by the geological time scale, is the entire distance from the tip of your fingers on the left hand to the tip of your fingers on the right. If someone were to run a nail file across the fingernail of your right middle finger—and that represented time—it would represent the amount of time humans have been on the planet!

What are the **time units** used in the **geological time scale**?

There are about six major time units in the geologic time scale. They are not precise, but are merely ways of trying to keep track of the Earth's long history. They are called eons, eras, periods, epochs, ages, and subages. The eon represents the longest geologic unit on the scale; on some scales, it is defined as a division of one billion years. An era is a division of time smaller than the eon, and is normally subdivided into two or more periods. The period is a subdivision of an era; the epoch a subdivision of a period; an age a subdivision of an epoch; and a subage (although it is not often used), the subdivision of an age.

What do the **divisions** on the **geologic time scale** represent?

The geologic time scale is not an arbitrary listing of the Earth's natural history—each specific boundary between divisions represents a change or event that delineates it from the other divisions. In the majority of cases, a boundary is drawn to represent a major catastrophe or a major evolutionary change in animals or plants (including the evolution of specific species) that lived during that time.

What is the **Phanerozoic eon**?

The Phanerozoic eon represents the time from 544 million years ago to the present—the time when the fossil record became much richer. The rough translation of Phanerozoic (a word of Greek origins) is "abundant life." The eon comprises the Paleozoic, Mesozoic, and Cenozoic eras.

How do scientists **divide the Precambrian eon**?

The Precambrian (4.55 billion to 544 million years ago) is broken down into various divisions, often depending on the country in which the scale was made. The following table shows two generally accepted versions of the Precambrian eon.

Divisions of the Precambrian—version I

Vendian	began 650 million years ago; ended 544 million years ago
Proterozoic	began 2.5 billion years ago; ended 650 million years ago
Archaean	began 3.8 billion years ago; ended 2.5 billion years ago
Hadean	began 4.55 billion years ago; ended 3.8 billion years ago

Divisions of the Precambrian—version II

Proterozoic	upper—began 900 million years ago; ended 544 million years ago
	middle—began 1.6 billion years ago; ended 900 million years ago
	lower—began 2.5 billion years ago; ended 1.6 billion years ago
Archean	began 4.55 billion years ago; ended 2.5 billion years ago

What does the **Precambrian eon** represent?

The Precambrian is the longest span of time on the geologic time scale, lasting from 4.55 billion years ago to 544 million years ago—about seven-eighths of the Earth's history. There are few fossil clues from this time; thus, generally speaking, scientists lump many events into this division. For example, this time span includes the formation of the Earth, the beginnings of the planet's crust, the formation of the first plates and their movements, the first life on the planet, and the evolution of an oxygen-rich atmosphere. At the very end of the Precambrian, the first multicelled organisms, including the first animals, evolved in the oceans; this event marks the beginning of the Paleozoic era—and of the Phanerozoic eon.

What does the **Paleozoic era** represent?

The Paleozoic era is delineated by two of the most important events that occurred in the animal world: At the start of this era, multicelled animals experienced an explosion in numbers and diversity (the event is called the Cambrian Explosion); within a few million years, almost all modern animal groups (phyla) appeared. For the duration of the Paleozoic era (which lasted about 300 million years), animals, plants, and fungi began to colonize the land, and insects began to occupy the air. Many of the periods within the Paleozoic era are loosely based on these events. The end of the Paleozoic era (and end of the Permian period) is defined by the largest mass extinction in the planet's history—an event called the Permian extinction. It wiped out the majority of organisms on Earth, including between 90 and 96 percent of all marine animal species.

What does the **Mesozoic era** represent?

The Mesozoic era began at the end of the Paleozoic era, and lasted for approximately 180 million years. It ended with another mass extinction that eliminated more than 50 percent of the species on the Earth, including the dinosaurs. According to the geological time scale, it was followed by the Cenozoic era.

The Mesozoic was a time when the planet's terrestrial plants changed dramatically. Early in the era, ferns, cycads, ginkgoes, and other unusual plants dominated; in the middle of the era, gymnosperms such as conifers (evergreens) became prolific; toward the end of the era, in the middle Cretaceous period, the earliest flowering plants (called angiosperms) took over from many other plants. This was also a time of great animal diversification: amphibians evolved; then the reptiles evolved from the amphibians; and then mammal-like reptiles evolved from reptiles.

During the Mesozoic era, the dinosaurs became the dominate forms of animal life on land. In the oceans reptiles also dominated, including such creatures as ichthyosaurs and plesiosaurs.

What does the **Cenozoic era** represent?

The Cenozoic era is the most recent of all the eras, and has spanned only about 65 million years. It began with the mass extinction at the end of the Mesozoic era, and continues through today. This era is sometimes referred to as the Age of Mammals, but in reality, the name could also be the Age of the Birds, Fish, Insects, and Flowering Plants—as all these life forms have diversified and grown in number over the past 65 million years. The Cenozoic is grouped into two main divisions: The Tertiary and Quaternary periods. The present time is part of the Quaternary period.

What is **relative time**?

Relative time is used as a means of establishing rocks and fossils in a general chronological order. It is based on where a rock layer is located in comparison with other rock layers; thus, it is only relative, not absolute, time. In the nineteenth century, scientists used this method to date rock layers—and to establish and construct the first geologic time scale.

What is **absolute time**?

Absolute time is the determination of the (approximate) true age of the rock; that is, how long ago the rock layer formed. Absolute time is determined by radiometric means (radiometers are instruments that measure the amount of radiation in the layers of the Earth), and was used to add more precise time spans to the geologic time scale, which is a relative scale. The techniques to determine absolute time were not perfected until after the 1920s.

FOSSILS

How do scientists determine the **past history of the life** on our planet?

One of the most useful clues that scientists use to determine the history of our planet—both its life and its changes—are fossils. The fossil record,

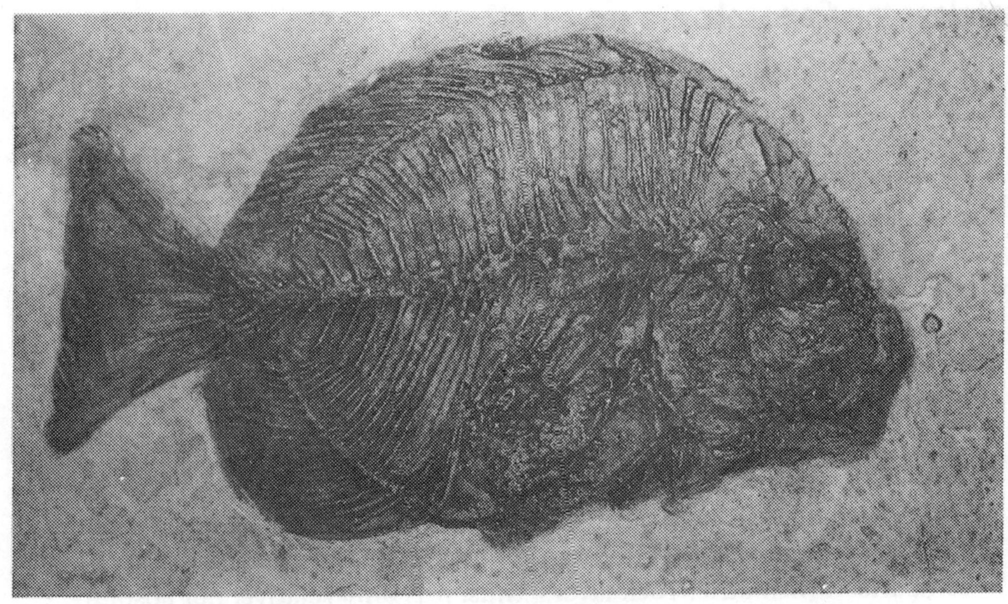

A 50-million-year-old imprint fossil of an extinct member of the spadefish family. It was found in Verona, Italy.
CORBIS/Sally A. Morgan; Ecoscene

preserved in rock layers over millions of years, allows us to understand our past—and where we are going in the future. Without fossils, our planet's immense and varied past would remain unknown to us.

What are **fossils**?

Fossils are the remains of plants and animals that have been preserved in the rock layers of the Earth. Fossils are often close to an organism's original shape. The word fossil comes from the Latin *fossilis,* meaning "dug up." There are many different types of fossils: The organism's remains and conditions present at the time the organism died determine what kind of fossil is formed. Most people are familiar with fossils that were formed in rock by the hard parts of an organism, such as teeth, shells, or bones, leaving an imprint. But animals and plants have also been preserved in other materials besides rock: Fossils can be found in ice, tar, peat, and the resin of ancient trees. The process of fossilization continues today—on land and in the oceans—as organisms die and are quickly buried.

How does a **fossil form**?

A fossil can form in a number of ways, depending on the type of remains and the environment in which the organism died. The general process is for the hard parts of animals (such as bones, teeth, and shells) or the seeds or woody parts of plants to be covered by sediment—sand or mud—either on land or on the ocean floor. Over millions of years, more and more layers of sediment accumulate, burying the remains of organisms deep within layers of sediment. The pressure of the overlying layers causes the sand or mud to eventually turn to stone; this process is called mineralization. The organisms' remains are often chemically altered by mineralization, becoming a form of stone themselves. The same process also produces petrified wood, coprolites (petrified excrement), molds, casts, and trace fossils.

Are **mineralization** and **fossilization** the same thing?

Not exactly. As far as paleontology is concerned, mineralization is a process of fossilization in which the organic compounds are replaced by inorganic material (minerals).

What are **molds** and **casts**?

Molds and casts are types of fossils—impressions of an animal's or plant's hard parts (and sometimes soft parts) that are left in the rock after burial and decay. Molds are hollow impressions (cavities) in the rock; if the mold is filled with sediment, it can often harden, forming a corresponding cast.

What are **trace fossils**?

Not all fossils are hardened bones, teeth, or shells—or even molds or casts. Some are trace fossils: Physical evidence, usually left in soft sediment (such as sand or mud), that creatures once crawled, walked, hopped, burrowed, or ran across the land. For example, small animals in search of food may have bored branching tunnels in the mud of the ocean floor; the tunnels were then filled in by sediment, buried by layers

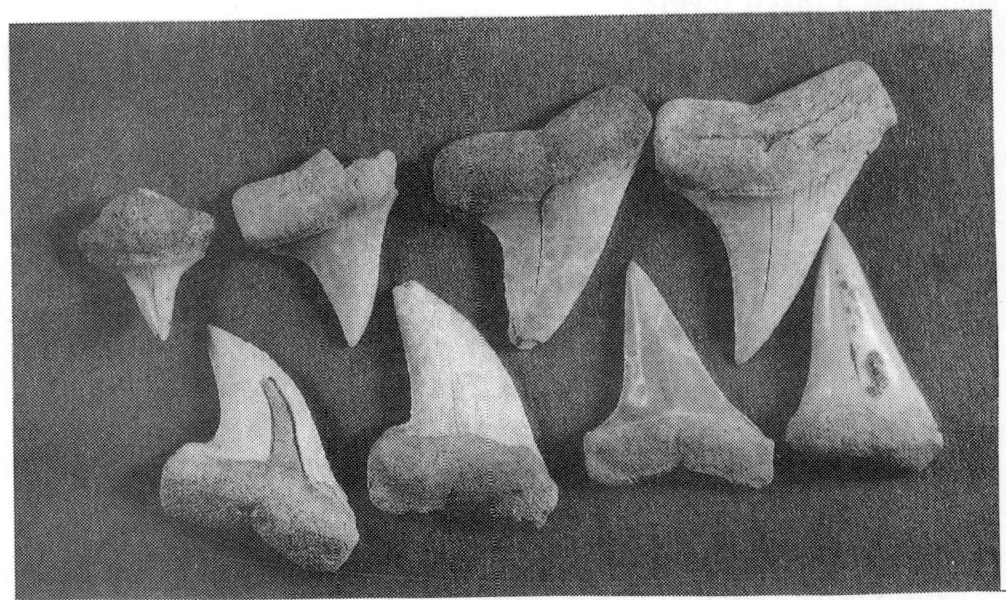

Animal hard parts (such as bones and teeth) can become fossilized, as were these eight shark teeth. *CORBIS*

of more sediment over millions of years, and eventually solidified, forming trace fossils. Trace fossils also include animal tracks; for example, when dinosaur footprints left in soft mud along a riverbank filled with sediment, the sediment eventually solidified in the form of the footprints. Today we see the results of this long-ago activity as trace fossils. Many originators of trace fossils (in other words, the animals that created them) are unidentifiable since sometimes there are no hard fossils of the creatures left in the area—just the traces of their passing.

What is **taphomony**?

Taphomony is the study of the way in which an organism was buried and of the origin of plant and animal remains. Through taphomony scientists try to "reconstruct" an animal or plant based on the evidence provided by its fossils. But, for the following reasons, this is a very difficult science that has many variables.

Scavenging and decay: After an animal's death, scavengers (birds and other animals) remove the soft flesh from the carcass. The hard parts (bones, teeth, or shells) that are not eaten begin to decay. The rate of decay varies by environment: Decay is faster in a humid climate and slower in an arid climate. The hard parts (the skeleton, shell, or the fibrous structure of the animal) are further reduced by the action of wind, water, sunlight, and chemicals in the surrounding environment. If the hard parts are completely worn away (through the action of wind, water, and sunlight) without being buried, no fossil can form. Often the fossil that forms is the result of a partial wearing-away of an animal's or plant's hard parts, making it difficult for scientists to reconstruct the life form that left the fossil.

Location: If an animal dies in a place where it is buried quickly, the organism has a good chance of surviving as a fossil. The ocean provides an ideal environment for fossilization—since organisms living here are very likely to be buried by sediment soon after they die. But if an animal dies in a vulnerable or quickly-changing environment, the bones and other hard parts can be broken or scattered. For example, the action of flash flooding can break up bones and, as the waters recede, the skeletal remains could be carried to various locations. However, flash flooding may also increase the chance of fossilization by moving the bones to a better area for preservation, such as a sandbank in a river.

Rapid burial: One of the best ways for a fossil to form is by rapid burial. If an organism's remains are quickly buried by mud or sand, the amount of oxygen is reduced, which slows the rate of decay. Again, the bottom of the ocean is an ideal place for this to occur. But even rapid burial does not ensure the creation of a fossil: As more and more sediment accumulates on top of the bones (or other hard parts of the organism), pressure from these overlying layers may cause damage to the organism's remains; acidic chemicals may also seep into the sediment, causing the bones (or other hard remains) to dissolve.

Fossilization: Another factor that figures into the creation of a fossil is a process called fossilization: The sediment (such as mud or sand) that surrounds the organism's remains must be turned into stone—which is accomplished by the pressure exerted by the overlying sediment; there must also be a consequent loss of water in the surrounding sediment. Eventually the sediment grains become cemented into the hard structure we call rock. As the spaces in the rock are filled

A cast fossil of a starfish (or sea star) from the mid-Devonian period, on display at the Black Hills Institute, Hill City, South Dakota. *CORBIS/Layne Kennedy*

with minerals (such as calcium carbonate or pyrite), the remains themselves may also recrystallize (forming a cast of the organism's remains).

Exposure: In order to study precious fossils, scientists must first find exposed rocks that contain them. (In other words, some of the fossil record remains buried—hidden from human view.) Fossils may be exposed by the uplift of land (which reveals fossil-containing sedimentary rock at the Earth's surface); by erosion of an area by wind, water, or even an earthquake; or via the development of land (for example, a road cut may expose a fossil-filled rock layer).

Why are there **gaps** in the **fossil record**?

Gaps in the fossil record are most often the result of erosion. Rock layers, including embedded fossils, can be eroded by the action of wind, water, and ice. Gaps in the fossil record can also be caused by mountain uplift (the process by which masses of land are "pushed up" to create mountains), which destroys fossils. Volcanic activity can also bury fossil evidence, as the hot magma rock physically changes the rock—and thus fossils—it touches.

How do we determine the **age** of **fossils**?

There are numerous dating techniques used to determine the age of rock and the fossils within the rock. Some of the more common methods use radiometric (isotopic) dating techniques. These techniques use the known rate of decay of radioactive elements into stable isotopes (atoms of a chemical element) or other elements to determine the age of the rock. With these techniques, fossils are not dated directly; the rock around the fossil is dated, and the fossil's age is extrapolated from the data.

Radioactivity within the Earth continuously bombards the atoms in minerals, exciting electrons that become trapped in the crystals' structures. Scientists often use two techniques called electron spin resonance and thermo-luminescence to determine the age of minerals. Both methods measure the number of excited electrons in minerals found in the rock: Spin resonance measures the amount of energy trapped in a crystal, while thermo-luminescence uses heat to free the trapped electrons.

After determining the number of excited electrons present in the minerals—and comparing it with known data representing the actual rate of increase of excited electrons in these minerals—scientists can calculate the time it took for the excited electrons to accumulate. In turn, the data can be used to determine the age of the rock and the fossils within the rock.

Another radiometric technique is uranium-series dating, which measures the amount of thorium-230 present in limestone deposits; these deposits form with uranium present, and have almost no thorium. Because scientists know the decay rate of uranium into thorium-230, the age of limestone rocks—and thus, the fossils found within the rock—can be calculated from the accumulated amount of thorium-230 found within a particular limestone layer.

What is **carbon dating**?

Carbon-dating techniques are used to determine the age of relatively young organisms (such as wood or bone) as well as the age of very ancient rock. To date younger organisms, the isotope carbon-14 (C-14) is used; it decays into nitrogen-14, and has a half-life (the time it takes for half the isotope to break down) of about 5,730 years. C-14 is produced in the Earth's atmosphere; it then combines with oxygen to form carbon dioxide. All organisms use this gas—as long as they are alive. When an organism dies, C-14 ceases to enter the organism, and the C-14 already present begins to decay. Scientists measure the amount of C-14 present and use its rate of decay to determine the age of the fossilized remains.

Scientists can also use carbon-dating techniques to determine if life existed in very ancient rock. Carbon has two other isotopes (atoms)—carbon-12 (C-12) and carbon-13 (C-13), the first slightly lighter than the second. Living organisms tend to "select" the lighter isotope, as it is easier to absorb (it takes less energy). When sedimentary rock is found to have more than the usual ratio of C-12 to C-13, scientists know that some form of life has altered the amounts—thus there was probably life present in the ancient sediment when it was deposited.

Why do **fossils form** more readily in the **ocean** than on land?

In the ocean, sediments are continually building up on the bottom, quickly covering the remains of dead organisms and preserving them. This greatly increases the chances of fossilization.

PALEOCEANOGRAPHY

What is **paleoceanography**?

Paleoceanography is the study of ancient oceans; it is an interdisciplinary field of study that uses data from geology, biology, physics, and chemistry, among other disciplines. Paleoceanographers can be thought of as detectives, piecing together clues from all types of samples, sources, and data, then using this information to reconstruct the events of the past. One of the central goals of paleoceanography is to reconstruct how the oceans have changed over time. Much of the data is gleaned from ocean sediment samples and is placed in databanks—where it can be accessed and studied by paleoceanographers worldwide.

How do paleoceanographers know that **oceans** have **changed over time**?

Paleoceanographers know that oceans—and the climate—have changed by studying sediments, including those collected from the deep sea. Over time, each layer of the ocean floor is laid down, one layer on top of the other. Cores drilled deep into the ocean floor show these sediment layers—all of which read like the pages of a book. By knowing which types of sediments formed when, and by analyzing some of the minerals

The *Glomar Challenger* was launched in the mid-1960s. It was the first research vessel to drill into the deep-ocean floor—bringing up invaluable information about the Earth's history. *CORBIS*

or fossils within the sediments, scientists have pieced together a good deal of information about how the ocean has changed over time.

The study of ancient oceans—and how they were created and evolved over millions of years—also gives us an idea of the physical mechanisms at work on our planet today. Paleoceanographers determine how the morphology (land configuration) of the Earth changed in the past, what caused these changes, and what effect such changes had on the oceans, climate, weather, plants, and animals. In this way, scientists can get a feel for what changes will occur in the future—and how those changes will impact our own species, *Homo sapiens sapiens.*

How do paleoceanographers gather **data about the past**?

Paleoceanographers obtain marine sediment and rock samples from the seafloor to decipher the ocean's past. They also rely on the interpretation of terrestrial rock layers that were once part of shallow and deep oceans.

How are **sediments obtained** from the **deep-ocean floor**?

Obtaining sediments from the deep-ocean floor is only a recent development. In the mid-1960s, a special ship fitted with a drilling rig, the *Glomar Challenger* was launched; it was named after the company that built it (the Global Marine Company) and after the HMS *Challenger*—the first true ocean research vessel, which sailed the oceans in the 1870s. The *Glomar Challenger* was the first to drill into the deep-ocean floor, bringing up long cylinders of sediment called cores from the bottom of the South Atlantic; the drills brought up material extending about 328 feet (100 meters) into the ocean floor. Improvements in drilling techniques have enabled researchers to penetrate more than 5,000 feet (1,500 meters) into the ocean bottom sediments, bringing up material representing up to 180 million years of the Earth's history.

What is the **composition** of **ocean sediments**?

There is no way to neatly categorize the composition of ocean sediments. But generally speaking, ocean sediments consist of a multitude of rock sediments and minerals, such as grains of sand, mud, silt, gold (from river deposits), heavy metals, and even ferromanganese (iron-manganese) modules—all depending on where the sediment is deposited. Additionally, sediment can contain small organisms, such as radiolarians (marine protozoans), diatoms (a type of algae), and calcareous nannofossils (minute fossils containing calcium). Material from larger organisms, such as the teeth of fish (such as sharks), corals, broken shells, whole skeletons, and sundry other marine organism remains can also be found in sediment layers.

What do ocean sediments tell scientists about the conditions in the early oceans?

Ocean sediments containing certain fossils, or of a certain composition, often give scientists clues about the makeup of early oceans—and even information on ancient biological communities. For example, any microfossils in the sediment samples that formed shells of calcium carbonate ($CaCO_3$) are good indicators: The relative amounts of light and heavy isotopes (atoms of a chemical element) are analyzed to obtain information about conditions in the ancient oceans such as the temperature, salinity, nutrient levels—and even to determine the volume of the ice sheets on land. This information can then be integrated with other data, such as paleontological, geochemical, and geophysical findings—and used to reconstruct ancient oceanic circulation patterns and global climate.

How are the ages of **ocean sediments** determined?

After the sediment samples have been recovered from the ocean floor—by means of a variety of techniques such as drilling or coring—they are prepared in the laboratory, where they are analyzed and their age is determined using techniques such as magnetostratigraphy and fossil identification.

In magnetostratigraphy, iron particles in the sediments are examined to determine their orientation to the Earth's magnetic field. As the sediments were laid down, the iron particles present were oriented with the magnetic field of the Earth at that time. A record of the reversals in the Earth's magnetic field are therefore preserved in the sediment layers by the orientation pattern of their iron particles. This pattern can be matched to magnetic reversal patterns found in the ocean floor basalts—rocks that can be dated by radiometric means (radiometers are instruments that measure the amount of radiation in the layers of the Earth).

Another method of dating ocean sediments is by the fossils and microfossils found in the various rock layers. Scientists use "marker species" found within the core—fossils in the sediments or rocks that have already been dated and the ages determined—to resolve the age of the sediment.

Are certain types of **ocean sediments** indicative of the **climate**?

Yes, scientists have found that certain sediments are often representative of certain climates. The following list cites the types of sediments that are often used to help determine past climates.

Ice-rafted sediments (tills): These sediments are indicative of glaciers and of a very cold climate.

Organic-rich sediments or phosphorites: These sediments are indicators of high (or low) productivity levels in the oceans—and could translate into the conditions of the climate.

Coral reefs: Coral reefs are indicators of past changes in sea level, since as the sea levels rise and fall, the coral reefs continue to follow the coastlines.

Pollen within sediments: In marine sediments, pollens from any vegetation can be used to determine the nature of the climate and changes in vegetation on land.

Eolian sediments: Eolian, or wind-blown deposits such as quartz grains, are often indicative of wind intensity and direction.

Fossil distribution: The way fossils are distributed in ocean sediments is one of the best indicators of climate. For example, plankton are sensitive to temperature: siliceous organisms typically represent cold climates and carbonates represent warm. The abundance of specific species of Foraminifera or Radiolaria can indicate the climate of a certain locale at a certain time; the discovery of cold (for example pachyderma) and warm species (such as sacculifer) in ocean sediments provide scientists with important clues about ancient climates.

Fossil chemistry: The details of the fossil chemistry are also strong indicators of ancient climate conditions. For example, fossil shells record the chemistry of the water; the ratio of stable oxygen isotopes (the oxygen-18 and oxygen-16 atoms) in seawater changes as the volume of ice in the oceans changes.

THE EARLY OCEANS

When did the **Earth first form**?

Scientists believe that the Earth formed about 4.55 billion years ago, with the Earth's crust becoming somewhat stable by about 3.9 billion years ago.

Is there a **connection** between the **Pacific Ocean and the Moon**?

The theory that the Moon came from the Earth was once thought to be foolish by most scientists. This idea was first proposed by British mathematician and astronomer George Darwin (1845–1912) in the early twentieth century. He called it the fission theory, in which a chunk of the rapidly spinning Earth was thrown off into space, and settled into orbit around the Earth. The connection between the ocean and the Moon was simple: The Moon was thought to have been thrown out from the area of the Earth we now call the Pacific Ocean.

The more accepted idea of lunar formation is that the Earth and Moon formed about 4.6 million years ago—as the material from a solar nebula (a collection of interstellar dust and gases) condensed to form our solar system. But amazingly, recent computer models have shown that the Moon may have truly come from the Earth. The theory is that a Mars-sized object struck our planet early after its formation. As the huge object struck the Earth, it tore material from far into the mantle (the part of the Earth that lies below the crust and above the core); this material eventually settled into orbit around our Earth. But the connection to the Pacific Ocean is not part of this more recent theory.

How did the **first ocean basins form**?

Although it is thought that some of the first "basins" were deep-impact craters on the early Earth's surface, the true ocean basins owe their origins to plate tectonics: The moving crust, driven by the movement of the Earth's mantle (the part of the Earth that lies below the crust and above the core and which is comprised of unconsolidated material), caused the separation of two types of material near the Earth's surface—the lighter and less-dense granitic rock that makes up the continents separated from the heavier and denser basalts that make up the ocean floor.

Where did **ocean waters come from**?

Scientists believe the Earth had two primary sources of water. First, the gases released from volcanic vents (in a process called "out-gassing") contained water vapor, which created clouds and eventually rain. Second, small (about 30 feet, or 9 meters, in diameter) ice comets, and perhaps frozen asteroid-type bodies, collided with the Earth, providing water for its basins. It is also known that by about 4 billion years ago, a half billion years after the Earth's formation, the planet's surface cooled enough for water to exist primarily as a liquid.

Have there been any **major changes** in the **oceans** over time?

The composition of the ocean's seawater, including the salts and trace minerals, has remained relatively the same over the Earth's long history. But there have been other major changes: About 2.1 billion years ago, as the planet's atmosphere began to accumulate more oxygen, so did its oceans. And of course, ocean organisms increased in abundance and diversity over time due to a number of factors—especially this increase in oxygen.

Why have **ocean shapes changed** over time?

The Earth's geologic activity is the main reason ocean shapes have changed over millions of years. Huge continental plates moved (and continue to move) across the planet, resulting in continental drift and plate tectonics (the movement of plates that comprise the Earth's surface). This activity not only changed the position of the continents, but also the shapes, depths, and bottom terrain of the oceans.

What was the **ocean that covered the entire world**?

Scientists believe that the largest ocean in our planet's history—or the ocean that covered the entire world—formed about 700 million years ago, toward the end of the Precambrian eon and not long before the beginning of the Paleozoic era (which began about 544 million years ago). This ocean, called the Iapetus Ocean, had only one major landmass, referred to as a "supercontinent"; it was called Rodinia.

What were the **continents and oceans** like during the **Paleozoic era**?

Early in the Paleozoic era, which began 544 million years ago, the supercontinent of Rodinia had already started to break up—forming the southern continent of Gondwanaland, a landmass consisting of parts of Australia, Antarctica, Africa, and South America, plus the Indian subcontinent (all south of the equator); and a northern continent (consisting of present-day North America) as well as some isolated landmasses north of the equator. The major ocean during the Paleozoic was the Iapetus Ocean.

By late in the Paleozoic, the scattered landmasses had joined into two large landmasses—Gondwanaland to the south of the equator and Laurasia to the north of the equator. Toward the end of the Paleozoic, these two continents slowly collided, forming the supercontinent of Pangea, meaning "all Earth."

What **major catastrophe** happened in the **oceans** at the end of the Paleozoic era?

At the end of the Paleozoic era (or the end of the Permian period, about 245 million years ago), almost all ocean life died out—that is, it became extinct. No one knows why, but there are theories that try to explain this massive global extinction. One theory includes evidence of oxygen starvation in the seas: With very little oxygen around, most living creatures in the seas died. The reason oxygen levels dropped is not known: Some scientists believe massive amounts of carbon dioxide from erupting volcanoes in Siberia warmed the globe, reducing the temperature differences between the poles and the equator. This warming slowed the ocean currents—causing the water to virtually stagnate.

Still another theory of why so many marine creatures died posits that a huge asteroid or comet struck the Earth, wiping out many species. But again, no one can explain why some species did not die, while others did—or where the comet impact occurred.

What were the **continents and oceans** like during the **Mesozoic era**?

At the beginning of the Mesozoic era (245 million years ago), there was essentially one large expanse of water called the Panthalassa Ocean (today's Pacific Ocean is the remnant of this huge ocean). This ocean surrounded the supercontinent of Pangea, meaning "all Earth." This giant landmass straddled the planet's equator roughly in the form of a **C**; the smaller body of water enclosed by the **C** on the east was known as the Tethys Ocean (or Sea). Only a few scattered bits of continental crust were not attached to Pangea, and lay to the east of the large continent. In addition, the sea level was low, and there was no ice at the polar regions.

Pangea broke apart about 200 million years ago (during the Jurassic period)—splitting again into the great supercontinents of Laurasia (to the north of the equator) and Gondwanaland (to the south). This breakup also saw the initial (east-west) separation of Europe from North America, and of Africa from South America. Thus the North Atlantic Ocean began to open up. Major rises in sea levels flooded many areas, including large parts of what are now Europe and central Asia.

What **modern continents** were represented in **Laurasia and Gondwanaland**?

The supercontinents of Laurasia and Gondwanaland can be translated into today's continents as follows: Laurasia included North America and

Eurasia (Europe, Siberia, and parts of eastern Asia), and Gondwanaland included South America, Africa, India, Antarctica, and Australia.

What is the difference between **Gondwanaland** and **Gondwana**?

There is no difference between the terms Gondwanaland and Gondwana. They are synonymous, and the use of the terms appears to be a personal preference.

What was the **Tethys Ocean**?

The large ocean called the Tethys Ocean (or Tethys Sea) formed about 245 million years ago, during the early Mesozoic era, when the single supercontinent Pangea began to split apart into two separate continents—Laurasia to the north and Gondwanaland to the south of the equator. As the two huge continents split, the gap between them slowly filled with seawater. In the beginning, the Tethys Ocean was very long, narrow, and shallow—but as the continents continued their movements away from each other, the sea became wider and deeper.

The Tethys Ocean was oriented roughly in an east-west direction, and was located just north of the equator for most of the Mesozoic era. This sea was the only body of water on the planet that was distinct from the Panthalassa Ocean, the very large ocean that covered much of the Earth.

Where is the **seafloor** of the ancient **Tethys Ocean today**?

As the continental movements squeezed the Tethys Ocean into today's remnants (the Mediterranean, Black, Caspian, and Aral seas), the seafloor of this ancient body of water was deformed and uplifted; a process that can be demonstrated by pushing together the ends of a rug. This uplifting produced the European Alps, the Caucasus, and the Himalayas—mountains where fossils of ancient marine life from the Tethys Ocean can still be found. Italian scholar, artist, engineer, and inventor Leonardo da Vinci (1452–1519) was the first European to discover these fossils in the Alps. He correctly assumed how and why these fossil-filled marine rocks were found at such great heights.

Are there any remnants of the ancient Tethys Ocean still present today?

In a stunning reversal of fortunes, the same continental movements that produced the Tethys Ocean in the Mesozoic era have almost erased any traces of this vast sea today. As the landmasses of Africa and India moved northward during the Cenozoic era, they collided with Europe and Asia, respectively. This movement squeezed shut the Tethys Ocean and only small remnants remain today: the Mediterranean, Black, Caspian, and Aral seas.

What were the **continents and oceans** like during the **Cenozoic era**?

By the beginning of the Cenozoic era (65 million years ago), the major modern continents had taken shape and they continued to move toward their present positions. The Tethys Ocean was larger, and still separated the northern and southern continents. Laurasia continued to split apart; Gondwanaland no longer existed as South America and Africa had already separated. Australia and India began to move northward from around Antarctica. By about 50 million years ago, the continents were close to their present-day configuration, with India slamming into Asia to create the Himalayas. Also, the Atlantic Ocean continued to open (widen), as North America separated from Europe.

How will the **oceans** look in the distant **future**?

Though no one really knows for sure, there are some interesting guesses. As the seafloor of the Atlantic Ocean continues to spread, the Atlantic will open wider. In the Pacific, continental plates will continue to subduct (move underneath each other), shrinking the now-largest ocean. And the Red Sea, in eastern Africa's Great Rift Valley, may also develop into a smaller ocean basin, depending on how fast the plates in that location continue to split apart (this activity is similar to the long-ago separation of South America from Africa, and of North America from Europe, which created the Atlantic Ocean). But all these processes will take hundreds of thousands to millions of years to occur.

LIFE IN THE ANCIENT OCEANS

When did **life** first **appear** on Earth?

Because the fossil record is incomplete, and the time scale immense, the origins of life on early Earth are highly debated. What scientists do know is that life first appeared in the early oceans, not on land; they estimate that this life began approximately 3.5 to 3.7 billion years ago (during the Precambrian). It's possible that life may have started even earlier—about 4 billion years ago—but scientist currently lack the evidence to support this idea.

Why is our **knowledge** of **early life** so **limited**?

Although oceans preserve and fossilize organic remains quite efficiently, scientists still lack knowledge of early organisms because of the very nature of the ancient organisms themselves and because of the nature of the Earth. There are three main reasons for the lack of knowledge: First, it is difficult to find 4-billion-year-old rock that has not been changed by heat, pressure, or erosion. Second, because the single-celled organisms (early life forms) are so small, they are difficult to find in rock. And last, because the organisms consisted of soft parts only, they probably decayed after death—leaving no (fossilized) trace of their existence. Without a fossil record, many of the organisms that were present in the ancient oceans will never be known to us.

Did **life** on Earth **develop continuously**?

Probably not; which is to say, there were "false starts." According to some scientists, life on Earth had to start many times. They theorize that once life began—either around a volcanic ocean vent or in a shallow pond or sea—comets and asteroids bombarded the planet, stamping out the beginnings of life. This may have happened many times over the course of the early Earth—perhaps for millions of years. Eventually, life was able to survive, adapt, and evolve.

What was the **probable first step** toward **life** on Earth?

The first step toward life on Earth was probably the formation of complex organic molecules (amino acids) from simpler organic molecules. There were many possible origins of these complex organic molecules, ranging from lightning strikes in the primitive atmosphere binding molecules together, to organisms from space—introduced by bombardment of organic-rich comets. Whatever their origin, many scientists believe organic molecules were common and stable on Earth at least 4 billion years ago, due to the absence of large amounts of free oxygen (also called a slightly reducing atmosphere). At some point, these complex organic molecules developed the ability to self-replicate; they also began to synthesize proteins and eventually became compartmentalized, leading to the first cells.

What was the **possible progression of life** on Earth?

The progression of life on early Earth is hotly debated in the scientific community. What follows is one possible scenario. But it's important to note that the dates are approximate—and even disputed.

3.7 to 3.5 billion years ago: The first primitive cells evolve.

3.5 to 3.2 billion years ago: Primitive bacteria may have evolved around ancient volcanic hot springs.

3.45 to 3.55 billion years ago: Photosynthetic bacteria (cyanobacteria) evolve.

2.2 to 2.1 billion years ago: The Earth's atmosphere becomes sufficiently oxygen-rich to support an ozone layer; this layer protects early evolving organisms from the Sun's harmful rays.

2.1 billion years ago: The first cells begin to form in greater numbers and develop into simple single-celled (unicellular) organisms.

1.5 billion years ago: The first eukaryotic cells form: These organisms have a nucleus, complex internal structures, and are the precursors to protozoa, algae, and all multicellular life.

What were the **first single-celled organisms** in the ancient **oceans**?

The first single-celled organisms in the ancient oceans were most probably anaerobic heterotrophic bacteria: They did not require free oxygen (anaerobic) and they obtained their food from external sources (heterotrophic). The next type of cells to evolve were probably the anaerobic autotrophic bacteria: They did not require free oxygen (again anaerobic) and they made their own food using the energy from the surrounding environment (autotrophic), such as deep-sea vents. Another term for these organisms is chemoautotrophic, because they made use of chemical energy. But, as chemical energy waned, photoautotrophs developed, which made use of energy from light.

How were **marine organisms key** to the development of **life** on Earth?

Approximately 3.45 to 3.55 billion years ago (during the Precambrian eon), a certain microscopic organism evolved that would eventually change the world. Today, its living descendants are known as "blue-greens," or cyanobacteria—microorganisms that secrete lime and form stony cushions called stromatolites. Some of the oldest known fossil stromatolites are found at a place called North Pole, Australia; living ones still exist in such places as Shark Bay, Australia.

The blue-greens were photoautotrophs: They used photosynthesis to produce their own food by obtaining energy from the Sun's rays and hydrogen from water; as they removed the hydrogen, they released oxygen. Over time, cyanobacteria released large amounts of oxygen into the atmosphere, most of which became "locked up" by iron on land and in the oceans. Eventually, the amount of oxygen produced was enough to exceed the capture by iron—and by approximately 2.2 billion years ago, the early atmosphere became filled with oxygen.

Stromatolites are stony cushions formed by lime secreted by cyanobacteria. Fossilized stromatolites have provided scientists with important information about the development of life on Earth. ***NOAA/OAR National Undersea Research Program***

The presence of oxygen in the atmosphere spurred the evolution of organisms that used this gas; the first organisms to develop aerobic (oxygen-consuming) respiration soon appeared. This aerobic process was more efficient than anaerobic (oxygen-free) respiration—allowing the development of larger cells and multicellular organisms.

When did the **first soft-bodied animals** appear in the **oceans**?

Fossils show that the first soft-bodied marine animals appeared about 600 million years ago (during the Precambrian). They included a form of jellyfish, sea pens (anthozoans), and segmented worms.

When did **larger early marine animals** evolve?

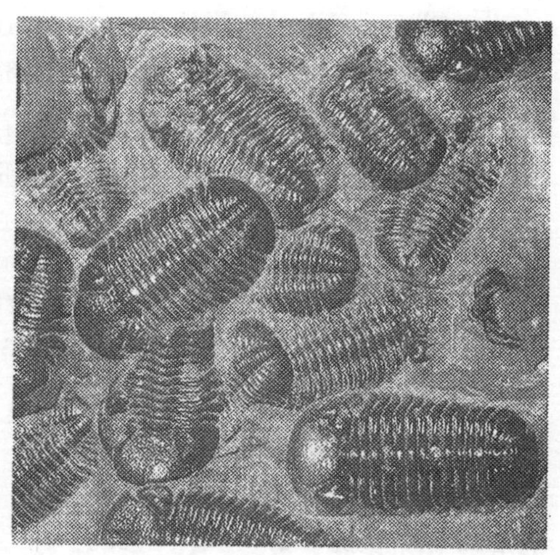

These fossils of trilobites, the first complex animals, are from the mid-Devonian period, or more than 360 million years old. *CORBIS/Kevin Schafer*

Some of the larger early marine animals evolved just after the end of the Precambrian eon, about 544 million years ago during the Cambrian period. At this time—nicknamed the "Cambrian Explosion" or "evolutionary big bang"—a great burst of evolutionary activity began in the world's oceans. New animals appeared rapidly (geologically speaking), filling the oceans with life. Some of these animals included trilobites (the first complex animals), sea pens (anthozoans), jellyfish, and worms.

No one really knows why the animals crowded the seas. One suggestion is that there was a radical change in the climate, allowing the animals to proliferate. Another theory suggests that an overall natural threshold was reached (for example, changes in temperature or oxygen levels), which provided an environment that was conducive to the proliferation of organisms. Whatever the reason, many of the sea creatures evolved from soft-bodied to having hard parts of some kind, including shells and internal and external skeletons. These changes allowed the sea creatures to better survive—and eventually leave evidence of their existence in fossils.

Why are the **Burgess shale marine fossils** so important?

One of the most significant records of life in the ancient oceans is found at the Burgess shale fossil site. Discovered in 1909 at Burgess Pass in British Columbia, Canada, the shale layer was formed from compressed muds laid down more than 500 million years ago. These muds are thought to have accumulated at the edge of an ancient ocean—and they preserve a variety of organisms that lived during the Cambrian period.

The muds and the nature of the location perfectly preserved the remains—making the Burgess fossils among the best in the world.

Many of the fossilized forms found in the Burgess shale can be recognized as ancestors of our modern-day marine animals, such as jellyfish and starfish. In addition, sea urchins, mollusks, echinoderms, ancient sponges, worms—and even arthropods (of which the trilobites are a member) were discovered there. To date, there have been more than 120 species of invertebrate animals (without a backbone) described from this one location.

What were the **first animals** to move **from the oceans to the land**?

The first animals to conquer the land may have been arthropods, such as scorpions and spiders. These creatures have been found in Silurian period rock layers, dating from 410 to 440 million years ago.

When did the **first plants move** from the **oceans to land**?

Fossils show that the first true land plants appeared about 420 million years ago (during the Silurian period), and included flowerless mosses, horsetails, and ferns. They reproduced by throwing out spores—minute organisms carrying the genetic blueprint for the plant.

Did any **dinosaurs** live in the **oceans**?

No, there were never any dinosaurs in the oceans; the true dinosaurs all lived on land. There were, however, other reptiles that once lived on land—but returned to the ocean to live and feed; these included the ichthyosaur and the plesiosaur. The ichthyosaur was small and had a dolphin-shaped body; the plesiosaur was large (nearly 50 feet, or 15 meters, long), with four paddle-shaped limbs and a long neck for catching fish. These reptiles became extinct approximately 65 million years ago, around the same time as the dinosaurs.

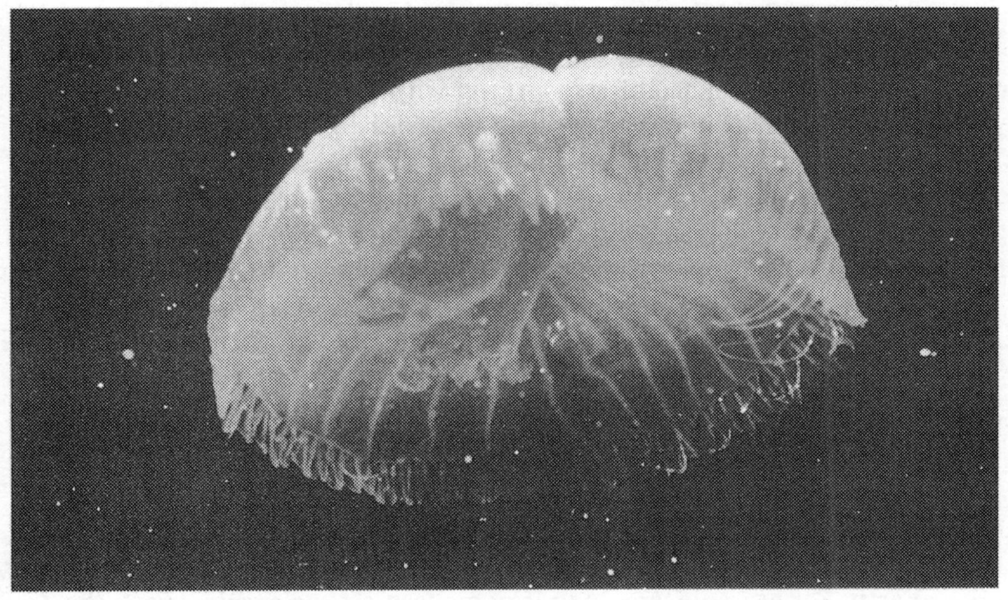

A form of jellyfish was among the first soft-bodied marine animals to appear in the oceans—about 600 million years ago.
NOAA/OAR National Undersea Research Program; M. Youngbluth

What were some of the **first marine mammals**?

Contrary to popular belief, marine mammals (such as whales and dolphins) did not evolve until a very short time ago—geologically speaking. Mammals evolved on land during the late Triassic period, about 208 million years ago; and they did not proliferate until after the end of the Cretaceous period 65 million years ago. It took even longer for mammals on land to move into the oceans—with some marine mammals evolving during the Miocene epoch about 23 million years ago.

Where are **prehistoric animals** being discovered below the **ocean surface**?

Not all the animal fossils found in the ocean sediments are from millions of years ago. Recently, some 60 feet (18 meters) below the ocean's surface off the coast of Georgia, scientists uncovered the remnants of prehistoric—about 14,000-year-old—animal life. During this time, the coastline extended out 60 miles (97 kilometers) beyond the present

Georgia shoreline, and the land was populated by enormous animals, such as bison, camels, and mastodons. The area is now completely underwater, and scientists working with Gray's Reef National Marine Sanctuary are uncovering fossils—not only of animals, but of the surrounding plants. So far, they have uncovered mastodon and bison bones, the tooth of a Pleistocene horse, and a marine worm burrow cast dated at about 18,000 years old. Plant fossils, such as pine pollen, and alder and grass seeds have been found—and may give scientists information on the ancient shorelines and possible changes in sea level.

What is a **living fossil**?

A living fossil is an animal or plant species nearly identical to organisms that lived millions of years ago. Many of these living fossils were first discovered as actual fossils before they were found living on our modern Earth. For example, one species of modern ginkgo survives from the Triassic period about 220 million years ago; the magnolia, one of the earliest true flowering plants, existed during the Cretaceous period, about 125 million years ago. Living fossils of animals include the tuatara, the only living survivor from a reptile group that was abundant during the Triassic period; and the coelacanth, a living fossil fish that was only recently discovered.

Another living fossil is the modern brachiopod *Lingula,* which is barely distinguishable from its Devonian period ancestor. One reason for the longevity of this species, or any other living fossil species, may be due to its ability to adapt and live in stable ecological niches. The *Lingula* live in the intertidal zone, a specialized niche along coastal areas. Even if the sea levels changed, these brachiopods could adapt by changing location to follow the water levels.

Why is the **coelacanth** famous?

The coelacanth (pronounced SEE-la-kanth) was a primitive fish that first appeared during the Devonian period of the Paleozoic era, almost 400 million years ago. Scientists believe this species of fish was the first to haul itself out of ancient oceans and onto land—and were probably the ancestors of many land animals. Numerous species of coelacanth

The coelacanth was thought to be extinct until one was caught by a fisherman off the coast of Madagascar in 1938. Others have been found since. Scientists believe this fish lives in the waters deep below the surface, which explains why they are rarely seen.

have been preserved in the fossil record, but all of them were thought to have gone extinct at the end of the Cretaceous period, about 65 million years ago.

However, in 1938, a fishing boat hauled up a living coelacanth from the waters off South Africa, with other specimens discovered over the years near the Comoros Islands near Madagascar. And in the late 1990s, another specimen was discovered off Indonesia.

The coelacanth is considered to be a living fossil by scientists and of great scientific interest. Fortunately, the modern species does not try to come onto land, where it would be vulnerable to predators and collectors. Instead, today's coelacanth appears to prefer a marine environment; the specimens from Madagascar and Indonesia appear to inhabit similar environments consisting of caves approximately 600 feet (183 meters) below the water's surface, situated along the steep sides of underwater volcanoes.

UNDERSTANDING THE OCEAN

THE INTERACTION OF OCEAN AND AIR

What are the features of **modern oceans**?

Modern oceans are similar to ancient oceans: They have strong currents, which are created by temperature differences (as the Sun does not heat surface waters uniformly) and by the mixing of equatorial (warm) and polar (cold) ocean waters; the oceans interact with the atmosphere; and, since about 600 million years ago, they have teemed with marine life—both fauna (animal life) and flora (vegetation). However, there are also differences between the oceans today and those of ancient times: The most significant difference is the diverse and plentiful marine life in the modern ocean. Another difference can be attributed to the Earth itself: Since our planet is tectonically active (its crustal plates are moving), the shapes and sizes of the oceans have changed over time. New oceans have been created, while others have diminished. It has only been within the last 50 million years that the continents and oceans shifted to their present positions.

How does the **atmosphere interact** with the **oceans**?

The atmosphere and the oceans could not exist without each other and they interact in a variety of ways to create currents, waves, climate, and weather. In simple terms, a region's climate is the result of the intense radiation from the Sun heating the oceans and atmosphere, creating oceanic and atmospheric circulations—patterns in the flow of water and air. Our weather systems are created by the localized rises and falls in the air, also governed by the oceans and atmosphere (warm air rises, while cool air falls).

How much of the **Sun's radiation** is **absorbed by the Earth's oceans**?

Over half of the Sun's radiation that reaches the Earth's surface is absorbed by the oceans. Some of this heat causes the evaporation of surface waters; the rest is stored in the ocean's surface layer or is moved downward into deeper waters by the dynamics of currents. The evaporated water ends up in the Earth's atmosphere mainly in the form of water vapor. Dissolved salts from the seawater also enter the atmosphere when the oceans are turbulent—for example, as sea spray during a storm. Each year an estimated 1 billion tons of sea salt particles enter the atmosphere from the ocean—and probably about 90 percent of these salts are eventually carried back to the oceans.

What is the **ocean's role** in the **water cycle**?

Evaporating water from the oceans and other waterways provides the water cycle (or hydrologic cycle) with vapor. The water cycle begins with the evaporation of water from the surface of oceans (and other waterways) and with evapotranspiration—the loss of water from soil by evaporation and by transpiration (through the membranes and pores) of vegetation. The evaporated water (now as water vapor) rises, reaching the upper atmosphere where it cools and condenses to create clouds. The clouds then produce precipitation. Some of the precipitation (in the form of rain, snow, sleet, etc.) falls into the ocean and adds to the seawater. Precipitation also falls on land, of course, where, if it is in the form of snow or ice, it can be stored on the ground as long as the weather conditions prevail to keep it frozen (but as soon as it melts, it becomes groundwater or runoff). Precipitation that seeps into the ground becomes part of the flow of groundwater, eventually reentering the atmosphere through evapotranspiration. Or, precipitation may remain on the surface (for example, when soil is already saturated or in highly developed areas), eventually running off into lakes, rivers, and streams.

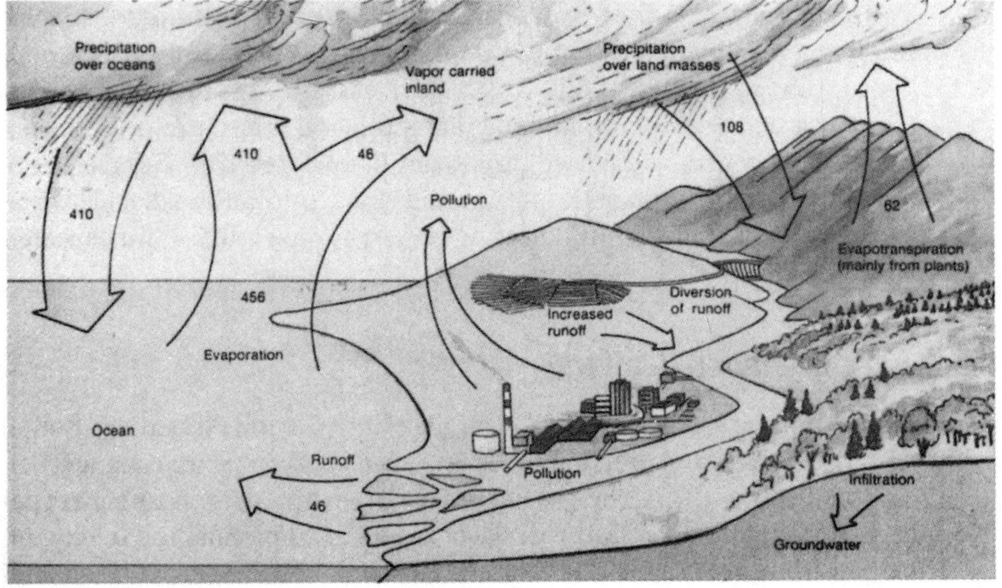

The circulation of water on Earth is shown in this diagram of the hydrologic cycle. *McGraw-Hill*

Eventually, all runoff returns to the oceans where it subsequently evaporates—and the cycle begins again.

What is the **ocean's role** in the **carbon cycle**?

The Earth's biosphere (the part of the world where life can exist) contains many carbon compounds—including natural gas, limestone, and the shells of certain marine organisms. Carbon is an essential part of the life cycle of all organisms—it is part of the processes of formation, transformation, and decomposition of fauna (animal life) and flora (vegetation).

The Earth's carbon cycle is complex. In very simple terms, carbon dioxide (CO_2) originates in volcanoes and carbonate rocks; it is then transmitted to living matter; and eventually, via metabolic processes in organisms, it is regenerated back to carbon dioxide.

The ocean's carbon cycle is somewhat self-contained: Seawater dissolves large quantities of existing carbon dioxide from the atmosphere. The gas also is released in certain processes in or around the oceans, such as carbon dioxide from volcanic eruptions or the dissolution of carbonate rocks.

The major players in the ocean's carbon cycle are organisms: Photoplankton (plant-like plankton that use photosynthesis) "fix" dissolved carbon dioxide during photosynthesis; oxygen is released, which then dissolves in the seawater. Zooplankton (animal plankton) and other marine animals, such as fish, consume the fixed carbon dioxide, and use the oxygen for respiration. And finally, plants and animals degrade into carbon dioxide compounds after they die—thus, releasing carbon dioxide into the atmosphere.

What is the **greenhouse effect**?

The term refers to a scientific theory that the Earth is warming. Some scientists believe the greenhouse effect is caused by an increasing level of carbon dioxide (CO_2). The gas occurs naturally in our atmosphere: The carbon cycle begins with the decay of plants and animals, the release of carbon from the oceans (for example, through volcanic eruptions), and through animal respiration (breathing). But CO_2 can also be produced by humans—through the burning of wood and fossil fuels (such as petroleum). Some scientists are concerned that this human production of CO_2 could throw off the balance of the gas in the Earth's atmosphere, resulting in an increase in carbon dioxide that would cause the atmosphere to trap heat that would otherwise be radiated back from the Earth—in the same way that the glass of a greenhouse retains heat from the Sun's radiation. Thus, the atmosphere's temperature would increase.

If the atmosphere's temperature were to increase, the seawater's temperatures would also increase—which would cause the water to expand, possibly raising sea levels around the world. In addition, the surface layer of the oceans would not be able to assimilate the increase in carbon dioxide, causing the oceans to become under-saturated with calcium carbonate. In the surface ocean waters, and especially the shallow seas, this would cause a major disruption of the carbon cycle, not to mention the circulation of carbon in general—and could have an enormous effect on world climate, marine life, and life in general.

What are **atmospheric cells** in the Earth's **atmosphere**?

Atmospheric cells are the major localized circulation features in the Earth's atmosphere—some of which were known even as far back as the eighteenth century. The cells are caused by the interaction of the air and oceans as each is heated by the Sun. The following list describes the Earth's three major atmospheric cells.

Hadley cells: As the rays of the Sun strike the Earth's surface (land and ocean) at the equator, humid air rises, creating a low-pressure system. As it rises, it reaches into the troposphere (the lowest layer of the atmosphere) and "splits," spreading north in the Northern Hemisphere and south in the Southern Hemisphere. The air then begins to cool as it heads toward the respective poles. At about 30 degrees north and south latitude, the cooled air sinks toward the surface and heads back toward the equator. This creates the "circulation" of air called the Hadley cells, named after George Hadley (1685–1768), the English physicist and meteorologist who first described them in 1753.

The Hadley cell has a major effect on the equatorial region: As the heated air near the equator rises, it produces thick clouds and heavy rainfall—producing a region of tropical cyclones and rainforests. As the air sinks at 30 degrees latitude, it encounters greater pressures and heat—essentially wringing the moisture out of the clouds, creating a band of dry, sinking air. It is this arid air that maintains the dry conditions found on the Atacama, Australian, Sahara, and Kalahari deserts around these latitudes.

Ferrel cells: The Ferrel cells form because not all the air in the Hadley cells flow back toward the equator—some air continues toward the poles along the Earth's surface. As the air reaches about 60 degrees north and south latitude, it runs into the cooler, polar air in each hemisphere, causing the air to rise, some of which heads back toward 30 degrees latitude. These mid-latitude global wind cells are called the Ferrel cells, named after William Ferrel (1817–91), an American meteorologist who first described them in 1856.

Polar cells: Finally, some of the rising air at 60 degrees north and south latitude continues to flow toward the respective poles. As it cools, it sinks to the surface—creating Polar circulation cells (sometimes called Hadley polar cells).

Why are **winds deflected** in specific directions in the Northern and Southern Hemispheres?

The primary (north and south) winds generated by the atmospheric circulation cells are deflected, or flow, in specific directions in each hemisphere: The winds curve to the right (or to the east) in the Northern Hemisphere and to the left (west) in the Southern Hemisphere. This is because of the Coriolis effect. The phenomenon was discovered by French physicist Gaspard-Gustave Coriolis (1792–1843) in 1835; the first person to explain the resulting deflection of the global winds was American meteorologist William Ferrel (1817–91). The Coriolis effect (also called the Coriolis force, although it is not really a force) is the apparent movement of objects when observed from a rotating system. In this case, because the Earth is rotating to the east under the air of the atmosphere, it appears to us that the winds are deflected.

What are the **major wind bands** in the Earth's atmosphere?

Because of the Coriolis effect, scientists divide the winds in the major circulation cells into six major bands across the planet. From the pole to the equator the winds are labeled as follows (there is one of each type in the Northern and Southern Hemispheres).

Polar easterlies: The polar easterlies flow from the east, out of the polar regions; the border between these winds and the westerlies is called the polar front.

Westerlies: The westerlies (also called the "roaring forties" and "furious fifties" because of the latitudes where they prevail) flow out of the west. They interact with the cold polar easterlies and the warmer air from the trade winds—a combination that often produces severe storms.

Trade Winds: The trade winds (which flow out of the northeast in the Northern Hemisphere and out of the southeast in the Southern Hemisphere) were discovered by English chemist and physicist John Frederic Daniell (1790–1845) in 1823. They are very persistent winds, deviating little from their compass direction. Long ago, the merchant marine put these winds to good use, which lent them their name.

Do the Earth's features affect the winds?

Even though there are major circulation patterns and wind bands, the planet's features do influence global winds: Mountains, continents, and ocean currents influence surface wind patterns—and produce distinctive climate zones.

Are there **areas** between the major **wind bands**?

Yes, there are converging areas between the major wind bands (or atmospheric circulation cells). One of these areas, between the Northern and Southern Hemispheres' trade winds, is called the intertropical convergence zone (also called ITCZ). This narrow band of hot and humid air near the equator is punctuated with frequent thunderstorms and squally winds. But at certain times of the year, the trade winds do not come together into a convergence—creating what is called the doldrums. Early mariners labeled this condition the doldrums after sailing ships became caught in its calmness. Sailors could drift aimlessly for days—or even weeks—without any winds.

The bands of the doldrums, ITCZ, and trade winds all change position, swinging across the equator twice a year—accompanied by seasonal changes. For example, during the year, the ITCZ migrates north and south only a few degrees of latitude over the Pacific and Atlantic oceans; in contrast, it changes as much as 20 to 30 degrees of latitude over South America, Africa, and a large region of Southeast Asia and the Indian Ocean.

At higher latitudes, between the trade winds and westerlies, there is a belt called the subtropical high pressure zone, or the horse latitudes. And higher still, between the westerlies and the polar easterlies, there is also a converging area, in which mid-latitude cyclones or frontal depressions occur; this "boundary" is caused by the fierce battle between these two opposing winds.

How is the **strength of the wind measured** on the oceans?

The strength of the wind on the ocean is determined by a standardized, descriptive scale that was invented in 1806 by English naval office Sir Francis Beaufort (1774–1857). The following scale was originally used to determine the effect of wind on what was called a "man o' war" ship in full sail; it is still used today and is known as the Beaufort scale.

Beaufort Scale

Force	Name	Kilometers per Hour	Miles per Hour	Description
0	calm	0–2	0–1	flat, slack sails
1	light airs	2–6	1–3	wavelets
2	light breeze	7–11	4–7	small waves, sails filling
3	gentle breeze	12–18	8–12	distinct waves, filled sails
4	moderate	19–30	13–18	wave caps, breeze bending branches on land
5	fresh breeze	31–39	19–24	some spray
6	strong breeze	40–50	25–31	cresting waves, holding an umbrella is difficult
7	stiff breeze	51–62	32–38	high sea, walking is difficult
8	fresh gale	63–75	39–46	difficult to stand unaided
9	strong gale	76–88	47–54	rolling sea
10	storm	89–100	55–63	violent seas, high waves
11	hurricane-like	101–117	64–72	blown-foam-covered sea, storm, high sea, impeded visibility
12	cyclonic	118+	74+	hurricane or typhoon storm

How is **speed measured** at sea?

Nautical speed for ships (it is also used for airplanes) is measured in knots. One knot is equal to one nautical mile per hour; for example, a ship traveling at a 30-knot speed travels 30 nautical miles an hour. Navigators use the nautical mile because of its simple relationship to the degrees and minutes when measuring latitude and longitude. The international nautical mile equals one-sixtieth of one degree, or a minute of arc, of the Earth's circumference. Thus, the international nautical mile equals 1.151 statute miles (1.852 kilometers).

The work "knot" originated long ago when sailing ships carried a speed-measuring device called a log chip and line: A log attached to a rope was heaved overboard—attached to a line that contained knots spaced at intervals of 47 feet, 3 inches (14.4 meters). As the log pulled on the rope and the rope was continually let out, the speed was calculated as the number of knots, counted in a standardized time interval of 28 seconds. (An interval of 28 seconds is to one hour approximately what a distance of 47 feet 3 inches is to 6,077 feet or 1.151 statute miles.) Thus, if the log pulled out 5 knots of line in 28 seconds, the sailing ship was moving at 5 knots or 5 nautical miles an hour.

WEATHER ON THE OCEANS

How do **clouds** form over the **oceans**?

Clouds are visible manifestations of collections of water droplets or ice crystals suspended in the air. These droplets are not like the rain we see falling during a thunderstorm on a hot summer's day—they are more than a hundred times smaller. Clouds develop over the oceans or land in the same way: Warmer, humid air, usually from the surface of the oceans or ground heated by the Sun—rises. As it rises into the atmosphere, it is cooled to its dew point (the temperature at which air becomes saturated and can no longer hold any more water vapor); condensation occurs in the form of tiny water droplets, which are suspended in the atmosphere by natural updrafts—what we view as a cloud.

Clouds forming over the oceans—especially in the more hot, humid areas around the equator—have a better chance of growing than do clouds over land, as the oceans continually supply the atmosphere with moisture. This is one of the reasons that tropical storm systems forming around equatorial-warmed waters often gain tremendous strength and size.

What is a **tropical disturbance**?

A tropical disturbance is an area of low pressure along the tropics; the tropics are the lines of latitude at 23.5 degrees north (called the Tropic of Cancer) and 23.5 degrees south (called the Tropic of Capricorn) of the equator. These are the lines of latitude where the sun is directly overhead during the summer solstice in each hemisphere. Tropical disturbances occur when huge clouds in these regions produce large, clustering thunderstorms. These disturbances can move through the tropics for at least 24 hours, but the winds never create any spinning movements within the clouds. One of the more well-known sources of these disturbances is an area off the west coast of Africa: The disturbances that form here head toward the Caribbean Ocean during the Northern Hemisphere's summer months, when the ocean surface temperatures are warmer. They usually settle down and dissipate—but sometimes, they can be the precursors to tropical depressions.

What are **tropical depressions** and **tropical storms**?

Tropical depressions form when a tropical disturbance strengthens over warm water—of at least 80° F (27° C); a tropical depression has sustained surface winds of 38 miles (61 kilometers) per hour or less. As the warm air rises, it condenses and releases energy that fuels the storm system. If these storms form at least 5 degrees north or south of the equator, the storms start to spin—counterclockwise in the Northern Hemisphere and clockwise in the Southern Hemisphere.

If the sustained surface winds (taken as a one minute average) increase to between 39 and 74 miles (63 to 119 kilometers) per hour, and the storm continues to rotate, a tropical storm forms.

What is a **tropical cyclone**?

Tropical cyclones form as tropical storms continue to gather strength from the ocean's warm water. These large-scale circular windstorms travel in the tropics and subtropics, with a well-defined low pressure area called an eye; they have sustained winds (taken as a one minute average) of at least 74 miles (119 kilometers) per hour.

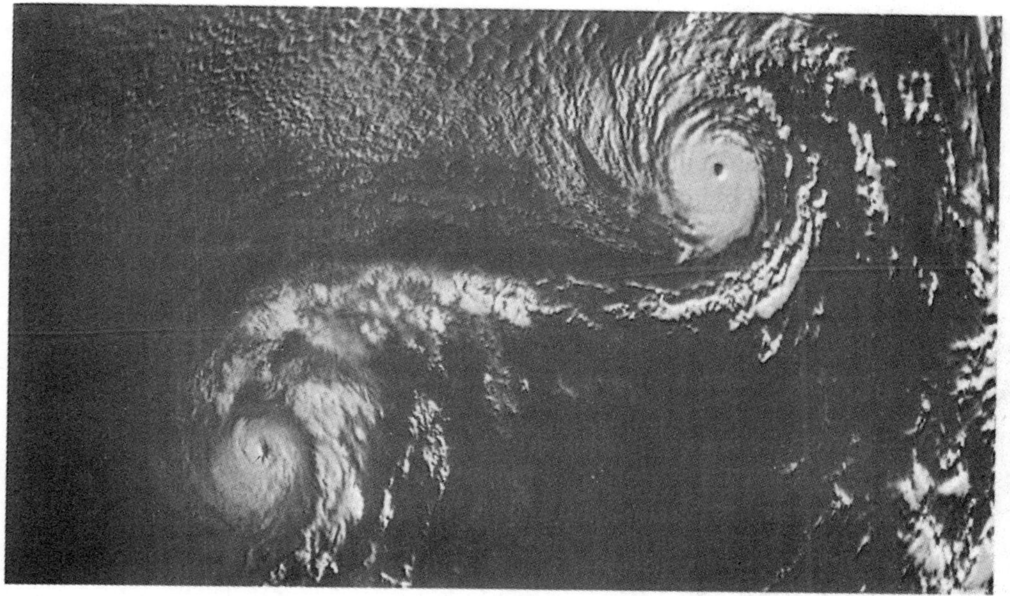

Satellite image of well-formed tropical cyclones Ione and Kristen in the Pacific Ocean, August 24, 1974. *NOAA*

What are the various **names** for **tropical cyclones** around the world?

Tropical cyclones around the world have a variety of names. For example, in the Indian Ocean, they are called cyclones; in Southeast Asia and the China Sea area, typhoons; off the coast of Australia, willy-willys; and in the Atlantic, Gulf of Mexico, Caribbean, and eastern Pacific, hurricanes.

How far **north or south** do **tropical storm systems** travel?

Tropical storm systems usually do not travel beyond 30 degrees north or south of the equator; any farther north or south of 30 degrees, there is not enough warm ocean water to either fuel or sustain the storm. In addition, these storms cannot form right at the equator, as there is no real spin of the winds here—and in fact, if such a storm crosses the equator, it would break apart. Of course, there are exceptions to the "warm waters only" rule. For example, tropical storms sometimes develop over the cooler ocean waters of the South Atlantic or southeastern Pacific Oceans.

What is a **hurricane**?

A hurricane is a tropical cyclone—a system that can become intense and long-lived if the conditions are right—and forms in the Atlantic Ocean, Gulf of Mexico, Caribbean Ocean, and east Pacific Ocean. Hurricanes have sustained winds (taken as a one minute average) of more than 74 miles (119 kilometers) per hour in any part of its weather system; the larger storms can have sustained winds of more than 150 miles (241 kilometers) per hour, with gusts up to 200 miles (322 kilometers) per hour. The storms have a clear, calm area at the center known as the eye, with bands that wrap around an eye extending up to 300 miles (483 kilometers) or more from the center. At the center, the air pressure is very low—the lower it drops, the stronger the winds around it and the stronger the storm.

Hurricane damage is caused not only by extremely high winds, but by torrential rains and storm surges (rises in sea level). Heavy rains can cause extensive flooding in coastal areas, and if a storm lingers, inland areas can also be flooded. If a storm lasts for several days and moves over great distances, the potential for destruction and death is immense. Storm surges can accompany a hurricane; these "bulges" in the water are created by winds and pressures from the system, which push ocean water before the advancing storm. The resulting "dome" of water (which can be as high as 25 feet, or 7.5 meters, above sea level) can be topped by wind-whipped waves. If a storm surge occurs during high tide, the damage to a shoreline can be even worse. Over time, hurricanes can also change the coastline by erosion of the beaches—not only during the storm, but from the accumulated effects of storms that occur close together or by many storms over time.

What was the **lowest pressure reading** ever recorded for an Atlantic **hurricane**?

So far, the lowest pressure reading for an Atlantic hurricane was 888 millibars, measured during Hurricane Gilbert in 1988. To compare, the usual reading for an average day at sea level is about 1013 millibars. (A millibar is one one-thousandth of a bar, a unit of pressure.)

Enhanced infrared imagery of Hurricane Hugo, centered off Puerto Rico. The category-4 storm later slammed into South Carolina, where the governor ordered a mandatory evacuation of coastal areas. *NOAA*

How is the **damage potential** of **hurricanes measured**?

Major hurricanes—of which there are approximately a dozen a year in the Atlantic Ocean—have the potential to destroy large areas of coastlines. For example, in October 1998, for a short period of time while over the Caribbean Ocean, Hurricane Mitch was classified as a category-5 hurricane, the most destructive known. Although Mitch struck Central America as a category-2 hurricane, its intensity left its scar on the land, as it dropped several feet of torrential rains on the region—causing more than $1.5 billion dollars in damage and killing more than 10,000 people.

Because of the danger to humans, a scale was developed to keep track of the intensity of hurricanes moving across ocean waters or land. The scale was proposed in 1971 by engineer Herbert Saffir and Dr. Robert Simpson, then the director of the National Hurricane Center. This scale includes wind speeds, estimates of barometric pressure at the hurricane's center, and height of a possible storm surge.

Saffir-Simpson Hurricane Damage Potential Scale

Category	Central Pressure millibars (inches)	Winds per Hour miles (kilometers)	Storm Surge feet (meters)
1 (minimal)	more than 980 (more than 28.94)	74–95 (119–153)	less than 6 (less than1.8)
2 (moderate)	965–979 (28.50–28.91)	96–110 (154–177)	6–8 (1.8–2.5)
3 (extensive)	945–964 (27.91–28.47)	111–130 (178–209)	9–12 (2.6–3.7)
4 (extreme)	920–944 (27.17–27.88)	131–155 (210–249)	13–18 (3.8–5.5)
5 (catastrophic)	less than 920 (less than 27.17)	greater than 155 (greater than 249)	greater than 18 (greater than 5.5)

What do **hurricane categories** mean in terms of **damage**?

Each hurricane category on the Saffir-Simpson Damage Potential Scale implies a certain amount of damage. The following list describes the kind of damage each level of storm is capable of producing.

Category 1 (minimal): The damage is primarily restricted to shrubbery, trees, and unanchored mobile homes. Other structures probably do not experience much damage. Signs (particularly those that are poorly constructed) may be damaged. Water inundates low-lying roads. Along the coast, there is minor damage to piers; smaller craft in exposed areas are torn from their moorings.

Category 2 (moderate): The damage includes considerable injury to shrubbery and tree foliage, with some trees blown down. There is also major damage to exposed mobile homes, some wreckage to doors, windows, and roofing materials (but not major destruction to buildings), and extensive damage to poorly constructed signs. Rising water about 2 to 4 hours before the hurricane makes landfall causes coastal roads and low-lying escape routes inland to be cut off. Piers are considerably damaged, and marinas flooded; small craft in protected areas are torn from their moorings. In preparation for a category-2 storm, evacuation of some shoreline residences and low-lying areas is required.

Category 3 (extensive): A category-3 hurricane causes foliage to be torn from trees, and larger trees to be blown down. There is some damage to roofing, windows, and doors, some structural devastation

The aftermath of the Galveston hurricane of September 1900, which devastated coastal Texas. The tropical storm claimed more than 6,000 lives, making it the deadliest hurricane to ever hit the United States. *NOAA*

to smaller buildings, and mobile homes are destroyed. There is also serious flooding along the coastline, with many small structures near the coast destroyed; larger coastal structures are damaged by the battering of the waves and by floating debris. Low-lying escape routes inland are cut off by rising waters about 3 to 5 hours before the hurricane makes landfall, and flat terrain that is less than 5 feet (1.5 meters) above sea level is flooded up to 8 miles (13 kilometers) inland. The evacuation of low-lying residences within several blocks of shoreline may be required.

Category 4 (extreme): Shrubs, trees, and all signs are blown down. There is also extensive damage to roofs, windows, and doors, with complete failure of roofs on many smaller residences. Mobile homes are demolished. Any flat terrain less than 10 feet (3 meters) above sea level is flooded inland as far as 6 miles (9.7 kilometers). There is major erosion of the beaches; flooding and battering by waves and floating debris cause major damage to the lower floors of structures near the shore. Low-lying escape routes inland are cut off by rising water about 3 to 5 hours before the hurricane makes landfall. In this case, massive evacuation of all residences within 1,500 feet (457 meters) of the shore may be required, as well as the evacuation of single-story residences in low ground within 2 miles (3.2 kilometers) of the shore.

Category 5 (catastrophic): Trees, shrubs, and all signs are blown down. There is considerable damage to roofs of buildings, with very severe and extensive damage to windows and doors; there is also damage to many residential and industrial buildings, with extensive shattering of glass in the windows and doors. Complete buildings are often destroyed, and smaller buildings are either overturned or blown away, with mobile homes demolished. Damage to lower floors is major, especially for those buildings that are situated less than 15 feet (4.6 meters) above sea level and within 1,500 feet (457 meters) of the shore. Low-lying escape routes inland would be cut off by the rising waters 3 to 5 hours before the hurricane makes landfall, and the beaches would be severely eroded. In this case, massive evacuation of low-lying residential areas within 5 to 10 miles (8 to 16 kilometers) from shore may be required.

How are **hurricanes named**?

People have named hurricanes for centuries. For example, hurricanes passing through the Caribbean Ocean were often named after the saint's day on which they occurred.

Before 1950, tropical storms were given a number based on when they occurred. The first storm of the season would be called "hurricane number 1"; the second, "hurricane number 2," and so on. For a short time, the military phonetic alphabet was also used as a naming scheme, so that hurricanes had names such as Able, Baker, and Charlie.

By 1953, all tropical storms were given women's names, with each successive hurricane assigned a name that represented the next letter in the alphabet. By 1978, both men's and women's names were used in the eastern North Pacific tropical storm lists; in the Atlantic basin, by 1979, the list expanded to include both male and female names.

More recently, the member nations of the World Meteorological Organization have revised the list to include names common to English, Span-

1992's Hurricane Andrew, a category-4 storm, devastated coastal areas. The storm's power is amply illustrated here—it was intense enough to toss these boats about like toys. ***NOAA***

ish, and French-speaking peoples. The order of women's and men's names alternate every year. For example, in 1995, the list of names began with Allison (which was followed by a man's name, Barry); in 1996, the names began with Arthur (which was followed by a woman's name, Bertha). There are six lists of tropical storm names, each comprised of 21 names from the letters A to Z, excluding the letters Q, U, X, Y, and Z. The lists are used on a rotating basis—thus, storm names are recycled.

When a tropical disturbance becomes a tropical storm, the National Hurricane Center in Florida gives the storm its assigned name. But if the storm becomes a hurricane that causes many deaths or extremely heavy damage, the name is retired. For example, on the Atlantic basin list, Andrew, Bob, Camille, David, Elena, Frederic, and Hugo are all retired names.

What do the various **weather watches** and **warnings** mean in coastal areas?

There are many watches and warnings issued by the weather services to help people understand the severity of a storm and take appropriate precautions. For example, a tropical storm watch is issued by the weather service when tropical storm conditions (winds from 39 to 73 miles per hour, or from 63 to 117 kilometers per hour) pose a possible threat to a specific coastal area within 36 hours; a tropical storm warning is issued when such conditions are expected in a specific coastal area within 24 hours.

In the case of a hurricane, a hurricane watch is issued when such a storm (or a developing hurricane) is a possible threat within 36 hours; a hurricane warning is issued if the same conditions are expected along a specific coastal area within 24 hours. In fact, a hurricane warning can remain in effect when the hurricane has caused dangerously high water, or a combination of high water and exceptionally high waves along the coast—even if the winds have calmed to below hurricane intensity.

What is a **monsoon**?

A monsoon is a seasonal, prevailing, and large-scale wind system caused by differences in the temperature between the land and the oceans. Monsoons, from the Arabic word for season (*mausim*), were the original names given to winds near Arabia that blew for six months from the southwest (during summer), then six months from the northeast (during winter). These winds also blow over India, southern Asia, northern Australia, and parts of Africa, and, though not as extreme, over North and South America. As the land and oceans "switch places" (in terms of which is warmer than the other), strong winds develop as cooler air rushes into the warmer areas.

Monsoons can be wet and dry. The wet monsoons usually occur in the summer and are accompanied by heavy rains—the moisture-laden winds flowing from the cooler sea to the warmer land. As the warm surface of the land heats the air, the air rises and the cool, moist air from the ocean moves in to take its place—creating the rains.

Dry monsoons have little rain, but still plenty of winds. They usually occur in the winter, with the winds flowing from the land to the sea: As the land quickly cools, it chills the air above its surface; the warmer air above the sea rises, allowing the cool, dry air from the land to push in and take its place, creating a dry monsoon.

What are **waterspouts**?

Vintage photograph shows the second of the so-called Great Waterspouts that occurred on August 19, 1896, in Vineyard Sound, off Martha's Vineyard, Massachusetts. These tall, whirling columns of air and water vapor can occur on the ocean as well as on lakes and rivers. *NOAA; Monthly Weather Review, July 1906*

Waterspouts are tall, whirling columns of air and water vapor that extend from a cloud to the surface of the ocean (they can also occur on a lake or river). Typically associated with normal rain showers and weak thunderstorms, a waterspout begins the opposite way from a tornado: A strong updraft of air is pulled upward; as the air rises, it rapidly rotates, producing an area of low pressure in its center. The moist air from the water rushes into the low pressure area, is rapidly cooled, condenses, and becomes visible. At the base of the column, seawater is sucked into the waterspout, rising only a few feet in the funnel; closer to the top, most of the water is fresh, formed by condensation.

Most waterspouts are more short-lived and less violent than are the majority of tornadoes. For people who happen to be in the ocean when a waterspout occurs nearby, the danger that is posed is from the winds that the waterspout generates; peak winds are typically between 50 and 100 miles per hour (80 and 160 kilometers per hour). Oceangoing vessels have also reported multiple waterspouts; and still other ships have reported a waterspout passing right over the vessel without causing any damage. Nevertheless, there have also been many reports of waterspouts blowing small vessels and people about, so caution is appropriate.

What are **land and sea breezes**?

If you've ever been near a sizeable body of water, such as a large river or lake, or an ocean, you've probably experienced land and sea breezes. The interaction of the land and the water are powerful: Sea breezes flow from the cooler waters toward the warming land during the day; land breezes flow from the cooling land to the warmer waters at night. It is

A beach at high tide. *JLM Visuals*

all a matter of air pressure caused by the difference in temperature: Warmer air is lighter and so it rises; cooler air is heavier and it "rushes in" to replace the rising warm air, causing the winds to blow.

TIDES

What are **tides**?

Each day, the surface of the ocean—and of those bodies of water that are connected to the ocean (such as gulfs, bays, and deltas)—rises and falls. Most ocean shorelines experience high tide and low tide every day. High tide is often referred to as "when the tide comes in"; and low tide as "when the tide goes out."

What **causes** the **tides**?

We would not know about tides—the natural up and down movement of the ocean surface—if we had no Moon. Tides are created by the gravita-

The same beach (as the one shown at left) at low tide. Note the measurable difference in the size and extent of the beach.
JLM Visuals

tional pull of our only natural satellite, the Moon, and to a lesser extent, our Sun. Because the Moon pulls on the Earth's oceans more than the Sun, the tides usually follow the Moon's cycle.

Tidal flows and cycles are also controlled by the friction of the water against the seafloor surface; the shape of the ocean basins; the presence of landmasses (the continents and islands); and sundry other complexities of ocean currents, winds, and interactions with the climate and weather.

What is the **tide-raising power** of the **Moon**?

The tide-raising power of the Moon on the oceans is 2.2 times the tidal influence of the Sun. It is estimated that the attraction the Moon exerts on a molecule of water on Earth is six million times smaller than the attraction from gravity—and of course, the Sun has an even lesser effect than the Moon.

What are the **spring and neap tides**?

The overall effect of the Moon and Sun on the Earth's tides changes over the course of a lunar month. The greatest effect on the Earth's tides results when the Moon and Sun reinforce each other's gravitational pull; spring tides are the result. These tides have nothing to do with spring—the term is from the German word, *springen,* "to jump." Spring tides occur when the Moon and Sun are in line with each other and are either on the same or opposite sides of the Earth. At such a time, the gravitational pull is the strongest and, therefore, the tidal range is the greatest. Neap tides occur when the Moon and Sun are at right angles to each other, with the pull the weakest and the tidal range the smallest.

What is a **tidal range**?

A tidal range is the difference in the height of the ocean between high tide and the next low tide. The average tidal range around the world is about 6 to 10 feet (2 to 3 meters).

What was the **greatest tidal range** ever recorded?

The greatest difference ever recorded between high and low tide was at Burntcoat Head, Nova Scotia, Canada. The tidal range measured 53.38 feet (16.27 meters) in the Bay of Fundy's Minas Basin.

What areas experience the **greatest tidal ranges**?

The greatest tidal ranges occur in bays and estuaries. The following list cites places where the average spring tide is in excess of 16 feet (5 meters).

Place	Average Tidal Range (feet / meters)
Burntcoat Head, Nova Scotia (Bay of Fundy), Canada	47.5 / 14.5
Rance Estuary, France	44.3 / 13.5
Anchorage, Alaska	29.6 / 9.0
Liverpool, England	27.1 / 8.3
St. John, New Brunswick, Canada	23.6 / 7.2

Place	Average Tidal Range (feet / meters)
Dover, England	18.6 / 5.7
Cherbourg, France	18.0 / 5.5
Antwerp, Belgium	17.8 / 5.4
Rangoon, Burma	17.0 / 5.2
Juneau, Alaska	16.6 / 5.1
Panama (Pacific side)	16.4 / 5.0

Where do **low tidal ranges** occur?

Some of the smallest tidal ranges occur in enclosed seas, such as the Mediterranean Sea, which is virtually tideless.

How many **tides** occur **per day**?

The number of tides in a tidal cycle is not consistent all over the world. The tides at any given location can be diurnal, semidiurnal, "mixed," or unequal:

diurnal: A diurnal tidal cycle is one in which there is just one high tide and one low tide per day. The Caribbean Ocean experiences a diurnal tide.

semidiurnal: A semidiurnal tidal cycle is one in which there are two high and two low tides per day. Certain places in southwest Florida experience semidiurnal tides.

mixed: A "mixed" tidal cycle is more complex; mixed cycles occur in areas of transition—which are affected by both semidiurnal and diurnal tides. Mixed cycles occur in northwest Florida's Apalachicola Bay (in the Gulf of Mexico) and in certain locations in the Pacific and Indian oceans.

unequal: Unequal tides are a condition between the diurnal and semidiurnal tidal cycles; in an unequal cycle, the two high tides are not at the same height on any given day.

How are **tides tracked**?

There are several ways to track tides. One way is through the use of tidal listings, which predict the time and height of a tide at a particular place. These listings are often used by shipping lines to determine the best times for shipping goods. Recreational sailors also use tidal listings to determine the flow of the currents along shorelines. The listings are based on years of careful records taken of past tidal performance in certain areas.

There are also tidal maps or charts. In one kind of chart, a cotidal chart, all points where high tides occur simultaneously are marked. Another type of cotidal chart connects points where the high tide comes up to the same height above sea level. On these charts, there is one point in a body of water in which all the radiating (or cotidal) lines meet; this is called the "no-tide" (amphidromic) point. There may be more than one such point in a larger body of water, such as the Caribbean Sea.

What is a **land tide**?

A land tide is just what it sounds like: In most areas, as the surface of the oceans rise and fall, the land responds, too—twice daily. In fact, the continental landmasses can rise and fall as much as 6 inches (15 centimeters) when the Moon is directly overhead and exerts its greatest pull.

What other effect do **tides** have on the **Earth**?

Tides have an additional effect on the Earth: They have a tendency to slow the planet's spin on its axis (rotation) by fractions of a second annually.

Do **tides** have an effect on **trees**?

Yes, scientists have found that trees bloat, and then shrink, with the rhythm of the tides. Measurement of the stems of young spruce trees show that their diameters change by several hundredths of a meter during a roughly 25-hour cycle. This cycle has two peaks, one higher than the other—similar to tidal cycles. These patterns may also explain the old folklore of cutting trees before a new moon to get the wood to dry

What is a storm surge?

A storm surge is an abnormal rise in the sea level along an open coast. This sudden rise in water usually occurs during a storm, as the onshore winds or lower atmospheric pressure pushes and "piles" the water up against the shore. If a high tide occurs at the same time, the storm surge adds to the height of the water—often causing devastation to a populated coast.

faster: New moons produce weaker coastal tides, and it corresponds that plants may take in less water at this time.

What is the **Earth's tidal bulge**?

The term "tidal bulge" describes the effect of the tide on the side of the Earth closest to the Moon—or where the lunar attraction is greatest.

Where do the **highest tides** occur?

The highest tides on Earth are found in Nova Scotia's Minas Basin, off the Bay of Fundy, where the water level at high tide can be as much as 52 feet (16 meters) higher than at low tide. The funnel-shaped Bay of Fundy squeezes and pushes the Atlantic Ocean tides to tremendous heights. These high tides occur every 12.5 hours, and an hour later every day, as they follow the changing position of the Moon. Amazingly, Nova Scotia "bends and tilts" when the tide comes in, as about 14 billion tons of seawater spread into the basin twice a day; but you can't really see the land move. In fact, scientists estimate that there is so much water at mid-tide, the flow in the Minas Channel north of Blomidon is equivalent to the combined flow of all the rivers and streams on Earth. And around mid-tide at Cape Split, you can hear the "voice of the Moon"—actually, the roar of the tidal currents as they run by.

The Monkton Tidal Bore occurs on the Pedicodiac River, which flows into Canada's Bay of Fundy (between New Brunswick and Nova Scotia): High tide causes ocean waters to surge upstream; the bore occurs at the point where these waters meet the outflowing river waters. *CORBIS*

What is a **tidal bore**?

A tidal bore is an abrupt front of high water, caused by the high tide coming in and surging up a river; on the surface of the water, a tidal bore looks like a slight wave trying to move against the flow of the river. In Nova Scotia's Minas Basin, tidal bores surge up several rivers that flow into the basin. These bores also occur in a few other rivers when the high tide begins, including Canada's St. Croix, Meander, Maccan, and Salmon rivers.

What is a **tidal inlet**?

A tidal inlet usually forms in gaps between barrier islands. Strong currents flow inward and outward through these gaps as the tide rises and falls. The inlets change over time, too: If a major storm reaches the area, it can cause a breach in the island, cutting another inlet; over time, moving sands also can close up a tidal inlet.

What can happen when the **tide turns**?

When a tide turns (from low to high or high to low), the action can create opposing currents in the water. Those currents meet, and often create a whirlpool, a swirling eddy of water. One of the most violent events that occurs when a tide turns is called a maelstrom, a feature that often forms in the Lofoten Islands off northern Norway.

Is it possible to harness **tidal energy**?

Yes, it is possible to harness tidal energy. This energy, produced as the ocean waters surge during the rise and fall of the tides, has long been a way to generate electricity. In the mid-1960s, the world's first plant run on tidal energy opened in France—a dam built in an area where the Rance River empties into the English Channel. Unlike regular hydropower (generated by rivers), the reversible turbines in the dam allowed electricity to be generated as the tides came in *and* out. Similar to river-generated hydropower, it is difficult to find sites where tidal power-generation is possible—thus, it is not as widely used as other sources of electric power.

WAVES

What is an **ocean wave**?

An ocean wave is the large-scale movement of water molecules. It appears as a ridge or swell on the surface of the ocean. The highest point of a wave is the crest; the lowest point is the trough. The distance between the crest and a neutral point (the water's "at-rest" level) is the amplitude; the distance between the crest and the trough is the wave height; and the distance between one crest and the next (or the distance between two consecutive troughs) is the wavelength. A wave period is the time, in seconds, between the passage of successive wave crests (or troughs) at a stationary point.

There are many different types of waves in the oceans, and they can be extremely large or very small, depending on their source and the area in which they occur. The smallest waves, called ripples, have crests less than

The anatomy of a wave: This shot shows a well-formed crest and trough; the distance between the crest and the trough is the wave height. ***CORBIS***

1 inch (2 centimeters) apart. Longer waves are subdivided into shallow-water waves and deep-water waves, depending on the depth of water through which they move. Shallow-water waves include tidal or river bores, in which the water particles move forward and backward in a horizontal plane, with the crests representing "crowds" of water particles. The troughs have very few water particles. Very violent storms at sea or underwater seismic activity will produce fast-moving deep-water waves.

What is **wave energy**?

The motion of the waves carries an enormous amount of energy; this energy is generated by the source of the waves, such as winds. In deep-water waves, this energy is expended in a lateral motion; in shallower waters, wave energy is released by the water crashing on the shore.

How do **waves develop** in the ocean?

Most waves on the surface of the ocean are caused by the action of winds; the height determined by the speed of the wind, the length of

The power of ocean waves is manifested here: A French merchant ship labors in heavy seas in the Bay of Biscay, as a huge wave, common in this area of the Atlantic, looms ahead. *NOAA; published in* Mariner's Weather Log, *Fall 1993*

time it blows (duration), and the distance over which it continually acts on the water's surface (called fetch). As wind blows across the surface of the water, it tries to "drag" the water with it. Since the water cannot move as fast as the air, it rises, but gravity pulls the water back. For each wind speed, there is a point at which the waves stop growing; this is because the energy is dissipated by breaking waves at the same rate as the wave energy is added by the wind.

Other types of waves include tide waves, caused by the gravitational attraction of the Moon and Sun; and internal waves, generated within the ocean by numerous causes, including wave interactions, atmospheric disturbances (such as storm surges), and earthquakes.

What is a **seiche**?

Seiche is the occasional, rhythmic rise and fall of water in a lagoon or bay. This phenomenon is not tidal; in most cases, coastal seiches are the result of interactions between the water and the air—long waves rolling in from the open ocean where they are generated by strong winds, large atmospheric disturbances, or even seismic activity on the coast or underwater. Seiches can be a positive factor along a coast, bringing much-needed nutrients and dissolved gases to the coastal habitat. They can also be very destructive, ramming boats into moorings as the waters rise and fall.

What is the **tallest open ocean wave** reported so far?

In 1933, the American tanker U.S.S. *Ramapo* was traveling in the Pacific Ocean from Manila, Philippines, to San Diego, California. One large ocean wave, measuring some 112 feet (34 meters) high, was reported during a 68-knot windstorm, but this was not officially recorded.

Have scientists noted any long-term **changes** in **wave height**?

Over the past 30 to 40 years, the waves in the North Atlantic Ocean have appeared to be getting higher. Less than a century ago, the average wave measured about 7 feet (2 meters); recently, the waves seem to have increased to about 9 to 10 feet (about 3 meters) on average. There is no explanation for the increase, although some scientists believe it may be that the waves have really not grown in height, but rather that the way they're measured today provides more accurate readings.

What do **swell** and **sea** mean?

Waves within the area of wave generation are called sea; a swell is a long, smooth (crestless), and sometimes massive wave or a succession of long waves, usually occurring outside the area in which they were generated and moving quickly away from the source. As waves move outside the area in which they were generated—such as at a point of a violent storm at sea—they sort themselves out. When wave crests momentarily coincide with each other, each adds its height to the overall wave crest; other times, crests may coincide with troughs, thus canceling each other out and leveling the sea surface. Storms off the Antarctic coast frequently manifest themselves on the beaches of California as long swells.

What are **rogue waves**?

Rogue waves (also called non-negotiable waves by sailors) are large waves that appear out of nowhere, occurring with little or no warning. Although they are found in all the oceans, certain areas of the globe seem to propagate more of these huge waves than other regions—such as the Cape of Good Hope off South Africa.

These waves are rare, with the majority seen and experienced by crews on ships crossing the oceans. For example, when the H.M.S. *Queen Mary* was pressed into World War II transport service, it was nearly capsized off Scotland by what her captain called, "one freak mountainous wave." And every year a few supertankers or large vessels traveling along a standard route from the Middle East to the United States or Europe experience major structural damage from rogue waves. In 1996, one such cargo ship, with 29 people onboard, sank after being hit by a giant wave.

There are many scientists who do not believe rogue waves exist. One reason is that it is difficult to determine what causes them—after all, not only do they appear out of nowhere, but observers (the witnesses to the event) are often swept away by them. Other scientists believe the gargantuan waves do exist and they suggest several reasons why they occur. The most recent theory, based on a mathematical model, shows that certain ocean currents or large eddies can occasionally concentrate an ocean swell and create a rogue wave. These currents or eddies act like an optical lens or magnifier, focusing the wave action to create huge crests—or the large troughs that mariners have often described as "holes in the sea." One such "focusing" current is the Agulhas Current, which skirts the South African continental shelf. There may be more such currents that form within or near the Gulf Stream in the North Atlantic Ocean, which would explain what the captain of the *Queen Mary* observed off the British Isles during wartime.

What is a **trapped fetch**?

As the pressure from a fast-moving line of thunderstorms essentially "pushes" the surface ocean water, many small waves get in step with each other and pile up to form one giant ripple, sometimes more than 70 feet (21 meters) tall. This is often called a "trapped fetch"—when a storm moves along with the waves it generates, adding energy to the moving waves.

Australian windsurfer Luke Hargreaves rides the break of Maui's "Jaws"—the monster waves that form off the island's north shore, making it an international surfing mecca. *CORBIS/John Carter*

What causes the **huge waves** seen near the **island of Maui**?

Off the coast of Maui, one of the Hawaiian Islands, there is an underwater ridge—produced by an ancient lava flow—that acts like a giant magnifying lens, bending and enlarging Pacific Ocean storm swells. These conditions often produce monster surfing waves as high as 70 feet (21 meters). The waves result from the unique shape of the underwater ridge—not from the effects of submerged cliffs, as some observers had previously assumed. This produces "Jaws," a surfing mecca on Maui's northern shore, near the town of Haiku.

At Maui's Jaws site, the ocean depth changes abruptly from 120 feet (37 meters) to just 30 feet (9 meters). Most of the time, the waves are not affected by the ridge. But when swells from stormy weather in the North Pacific Ocean are longer than about 1,000 feet (305 meters), they "trip" over the ridge, wrapping around it. As part of a storm swell passes over the ridge crest, it slows because water travels slower in shallow water; the other parts of the swell travel faster in deeper water, causing the wave to focus on the ridge in a process called refraction. The swell bends inward as it travels on either side of the ridge, focusing its energy on the center of the wave crest—to form the monster wave.

CURRENTS

What is a **current**?

A current is the large-scale movement of water in a specific direction. In the oceans, currents can be thought of as the Earth's circulatory system, moving vast quantities of waters of different temperatures around the world, creating climates, providing nutrients for marine life, and influencing weather patterns. They may form permanent circulation systems, as in the Atlantic and Pacific oceans or currents can be relatively short-lived, affecting only a limited area—especially along a coastline.

What is a **gyre**?

A gyre is the large-scale circular movement of currents—usually a permanent feature—on the surface of the ocean. Gyres normally occur when the surface current, pushed by the Earth's prevailing winds, runs up against a continent. This causes the current to modify its path into a giant, spinning circulation of water. The North Atlantic gyre is formed by the equatorial current, which flows west; it is forced by the North American continent to flow northeast along the coast of the United States as the Gulf Stream Current; this continues to the east toward northern Europe as the North Atlantic Current; then heads southward as the Canaries Current and back to the west, completing the circulation loop.

What are **tidal currents**?

Tidal currents, a type of subsurface current, are the large-scale horizontal movements of water caused by the gravitational interactions among the Earth, Moon, and Sun. They manifest themselves as localized ocean currents off continental and island shores; in the open ocean, they diminish to the point of being undetectable. The direction of tidal currents changes throughout the tidal cycle; they create a clockwise gyre (large-scale circular movement of upper ocean surface currents) in the Northern Hemisphere and a counterclockwise gyre in the Southern Hemisphere.

What are **non-tidal currents**?

Non-tidal currents are usually ongoing movements of water—permanent features. They occur at the surface and the subsurface. Non-tidal currents are the circulation system of the oceans; like the atmosphere, the oceanic circulation system is mostly driven by the Sun's radiation and the Earth's rotation (especially the Coriolis effect). Temperature and density differences in the water and temporary winds also drive non-tidal currents, including subsurface gradient currents (which are caused by differences in the density of seawater). Other factors can influence the formation and maintenance of localized non-tidal currents, such as the depth of the water, the size and location of the land, the underwater topography, activity of volcanoes, and discharge from river systems.

What are **surface currents**?

Surface currents in the ocean are the continuous movements of water found at the surface to just a few feet below. These currents are generated by the planet's prevailing winds—which are a direct result of the Sun's radiation and Earth's rotation. For example, equatorial winds cause westward flowing surface currents north and south of the equator. If there were no continents to redirect their flow, these equatorial currents would continue to flow around the world. They are considered the most important ocean currents in the world, as they circle the ocean basins to either side of the equator, and create specific climate conditions in certain areas.

How do the **wind** and **rotation** of the Earth create **surface currents**?

The force of sustained winds on the surface layer of water on the ocean causes it to move; this motion is transmitted down through successive layers, but the motion slows with increasing depth due to friction. This

Tapes from recording meters aboard the *Oceanographer* are read off the North Carolina coast in 1940. Because of continued observation and careful record-keeping, scientists have been able to provide the Navy, the merchant marine, and the oceangoing public with better data on the ocean's currents. *NOAA/C&GS Season's Report, Borden 1949*

movement of water is called a wind current, and in order to form, needs a steady wind lasting at least 12 hours.

But this current does not flow in the direction of the wind, which would seem logical. The rotation of the Earth deflects the motion toward the right in the Northern Hemisphere, and toward the left in the Southern, leading to a difference in direction between the winds and the wind-generated currents. This difference in direction can vary between 15 degrees in shallow coastal areas, to 45 degrees in the open ocean.

How do **surface currents** influence **climate**?

Some surface currents influence climate by their close proximity to land. For example, in the North Atlantic, the Gulf Stream is a fast-moving current carrying warm water in a narrow band from the eastern coast of Florida along the East Coast of the United States, past Newfoundland, eventually arriving at England, where it moderates the climate. To contrast, the cold Labrador Current, which runs southward along the eastern Canadian coast and dips below the Gulf Stream to reach the eastern seaboard of the United States, gives New York, for example, a cool climate—even though the state is at the same latitude as warm southern Italy.

What **separates** the **waters of the deep ocean** from the **surface ocean waters**?

There is a permanent thermal barrier that separates the warm surface waters from the relatively deep, cold waters; this barrier also prevents the surface and the deep waters from mixing. The surface waters—which receive their warmth from the Sun's radiation and have wind-driven currents—make up about 2 percent of the ocean's volume. The remaining volume is made up of the deep, cooler waters, with currents primarily driven by differences in temperature and salinity—that is, differences in density.

What are **subsurface currents**?

Subsurface currents in the ocean are those that occur below the domain of the surface currents. They are also referred to as deep-ocean currents. They are caused by density variations in ocean water, which is the result of differences in temperature and salinity. Water that is denser (heavier) will sink downward and move along the seafloor toward the equator—thus creating subsurface, or deep-ocean currents. The movement of these types of currents is called thermohaline circulation.

For example, deep beneath the warm, northeast-flowing Gulf Stream Current, a large cold current flows in the opposite direction, called the North Atlantic Deep Current. This is caused by very cold, salty—thus dense—water from around Greenland, sinking and flowing toward the equator; it is also pushed up against the western edge of the Atlantic Ocean's basin by the Earth's rotation. As it flows toward the equator at approximately 0.5 miles (0.8 kilometers) per hour, it hugs the ocean bottom, displacing the warmer water upward—and creates a subsurface current. When it reaches south of the equator, it is often undermined by another deep-ocean current: The Antarctic Bottom Current, which flows northward toward the equator—and under the North Atlantic Deep Current.

Who **first discovered** ocean **currents**?

Probably the first currents were discovered around an area that had plenty of famous civilizations—not to mention shipbuilders: The Mediterranean Sea. Ships were used by the Egyptians, Athenians, the "sea kings" of Crete, Phoenicians, Romans, and others to carry on trade. And although there is not a true written record of currents, many ships left the confines of the Mediterranean to travel the Atlantic Ocean, reaching as far as Iceland and perhaps beyond. Some of the earliest records of currents were recorded by the Polynesians more than 2,000 years ago in the Pacific Ocean . They used their knowledge of the weather, winds, migrating seabirds, and ocean currents to navigate between their widely separated islands.

Why is it difficult to **measure ocean currents**?

One of the major reasons it is difficult to measure ocean currents has to do with the size of the oceans: It is difficult to collect a massive amount of data over such a large area. In addition, the ocean floor has huge, river-like channels—similar to those on the land surface, but not as easy to find. The currents are often dragged through these elusive channels, changing the overall current flow.

How have **athletic shoes and rubber toys** been used to chart **ocean currents**?

In 1990 a Korean container ship bound for the United States ran into problems, accidentally dropping 80,000 Nike athletic shoes into the Pacific Ocean. Though oceanographers could not follow the sneakers across the ocean, they did retrieve many of the shoes on beaches in Alaska, Oregon, and Hawaii. The shoe landings confirmed and added to several computer models of Pacific Ocean currents.

Other accidental ocean research "tools" were some 29,000 bathtub toys. During a Pacific storm in January 1992, the toys (rubber turtles, frogs, ducks, and beavers), which were being transported aboard a container ship, were swept overboard near the International Dateline. Like the athletic shoes, the toys washed up on various shores, helping confirm and add data to computer models of the currents.

What are **convergences** and **divergences**?

A convergence is where different currents come together; in the majority of cases, this causes the water to sink. A divergence is where different currents pull apart, causing the water to rise. Often the best evidence of a convergence or divergence is to watch the debris on the oceans—logs, seaweed, garbage, and even shipwrecks can gather at these points. Because this floating material is found in the middle of the oceans, there is often a large concentration of birds and marine animals that frequent these areas.

What are some of the **major surface currents** in the oceans?

There are many surface currents in the Earth's oceans. They include the North Equatorial Current (warm), the Equatorial Countercurrent (warm), South Equatorial Current (warm), Oyashio Current (cold), Kuro Siwo Current (warm), California Current (cold), Labrador Current (cold), Gulf Stream Current (warm), Benguela Current (cold), Canaries Current (cold), Peru (Humboldt) Current (cold), North Atlantic Drift (warm), and the Antarctic Circumpolar Current (cold).

What is the **North Equatorial Current**?

The North Equatorial Current (or Equatorial Current) includes a warm, east-to-west-moving Pacific Ocean current and a warm east-to-west-moving Atlantic Ocean current driven by the northeast trade winds that blow over the tropical oceans of the Northern Hemisphere. It eventually splits in two—as the opposite-flowing current called the Equatorial Countercurrent and the Kuro Siwo Current of Japan.

What is the **Kuro Siwo Current**?

The Kuro Siwo ("black stream") Current is a strong, fast-moving warm current that flows northeastward from the northern Philippines, up the east coast of Japan. It is a continuation of the North Equatorial Current, and carries warmer waters into the North Pacific Ocean. The approximately 50-mile- (80- kilometer-) wide and 1,300-foot- (400-meter-) deep current actually appears dark from far away—thus its name.

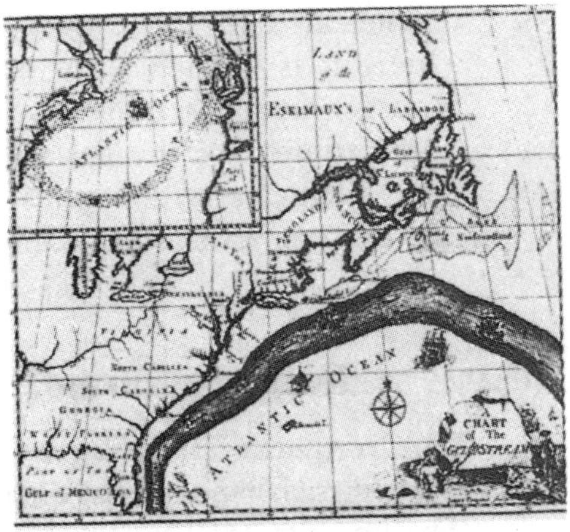

American statesman Benjamin Franklin drew this map of the Gulf Stream (c. 1782) to explain why ships returning across the Atlantic traveled slower if they dipped too far south, where their westward progress was impeded by the northeast-moving current of the Gulf Stream waters. *NOAA Central Library*

What is the **California Current**?

The cold, shallow California Current flows southeastward along the west coast of North America; it then meets the North Equatorial Current around the equator. The fog banks of San Francisco are caused by the California Current, as moist, relatively warmer air moves over the cold current's waters.

Why is the **Labrador Current** important?

The cold, slow-moving Labrador Current travels in a southward direction; its cold waters dip below the northeastward-flowing Gulf Stream Current. In the warmer, summer months in the Northern Hemisphere, the current often extends down to about Cape Cod, Massachusetts; in the winter months, it reaches down to about Virginia. The Labrador Current is also the current that carries icebergs into the major shipping lanes between North America and Europe during the winter.

Why is the **Gulf Stream Current** important?

The warm, fast-moving Gulf Stream Current is located just off the East Coast of the United States, and brings a temperate climate to the coastlines it passes; it is also part of the major Atlantic gyre (large-scale circular movement of currents). The Gulf Stream moves north from the

Caribbean Ocean; the warm water then reaches the Gulf of Mexico, where it is warmed even more. It then moves through the Florida Straits, where it becomes one of the fastest currents known. From there, it travels up the East Coast of the United States, traveling about 80 miles (130 kilometers) a day—until it reaches the colder Labrador Current off the coast of northeastern Canada.

What is the **Benguela Current**?

The Benguela Current is part of the South Atlantic Ocean gyre (large-scale circular movement of currents). It is a cold, slow-moving, broad, shallow current that flows northward along the west coast of South Africa. There is a great deal of upwelling ("pushing" of water) along this current, forming where currents part (diverge). These divergences provide food for many plants and animals, and are known for their dense marine life.

What is the **Canaries Current**?

The cool Canaries Current is a split off from the North Atlantic Drift, and tempers the climate of the Canary Islands—which would be tropical if the Canaries Current did not flow past the islands. As this cool current moves south, it meets the warm winds and waters of the Iberian peninsula, creating the famous fogs along the western edge of the Pyrenees Mountains between Spain and France.

What is the **Humboldt Current**?

The Humboldt Current, now called the Peru Current, is a cold, shallow, and slow-moving current that flows north and then northwest along the western coast of South America. It is a well-known area of upwelling (pushing of water caused by temperature differences or diverging currents), which encourages the growth of small plankton (nutrients). This, in turn, attracts a wealth of other marine animals that feed on the plankton, and other animals that feed on those animals. Thus, it is a tremendously productive fishing ground. If the Peru Current changes course, it inhibits the upwelling, affecting the marine life in the area. This change in course usually occurs during El Niño years, as warm waters move south, displacing the cold, northward flowing Peru Current.

Why is the **North Atlantic Drift** important?

The warm, shallow, and slow-moving North Atlantic Drift (or Current) is a branch of the Gulf Stream Current; it flows to the northeast, across the North Atlantic, and onward toward the Arctic Ocean. Because it is a split from the Gulf Stream, the warmer waters moderate the climate of western Europe and England, resulting in moderate winters for the latitude, and keeping Norwegian ports free of ice during the winter.

What are some of the **major subsurface currents** in the oceans?

There are many subsurface currents in the Earth's oceans. They include the Deep Western Boundary Current (cold), Cromwell Current (cold), Weddell Sea Bottom Water (cold), and North Atlantic Deep Water (cold).

What is the **Deep Western Boundary Current**?

The Deep Western Boundary Current is the world's largest deep-ocean current—a fast, cold current in the southwest Pacific Ocean. It is about 100 times the size of the Amazon River, and a part of the global ocean circulation system. In fact, this current channels about 40 percent of the world's cold, deep water throughout the oceans.

The current is also responsible for carrying a great deal of sediment into the oceans: As it travels north from the Antarctic Ocean to the Pacific, it hits the landmass of New Zealand, where the mountains provide an abundance of eroded sediment. Fine-grained muds drift into the current to form huge deep-sea drifts of sediment, some of which are several hundred miles long. In fact, scientists are using this buildup of sediments to find out how the climate has changed in the Southern Hemisphere and to determine the changes in the strength of the Deep Western Boundary Current over time.

What is a rip tide?

The so-called "rip tide," also called the undertow, is actually a strong, localized current that flows quickly outward from the shoreline. These currents form in a variety of ways: They can be caused by the rapid escape of water, by means of waves, from the shore through a gap in a reef; by waves that break obliquely (at an angle) across a longshore current (a current flowing parallel to the shore); by the head-on meeting of two tidal currents; or by a strong tidal stream that enters shallow water (a stream of moving water caused by tides that, for example, moves through an inlet). The rapid movements of rip currents return a substantial amount of water to the oceans and they often carry a great deal of sediment as they tear through the shoreline. They are also responsible for many deaths, as swimmers can become caught in the strong, moving waters, which carry them toward the open ocean.

What is the **Cromwell Current**?

Named after Townsend Cromwell (?–1958), the oceanographer who discovered it in 1952, this subsurface current flows swiftly from west to east along the equator in the Pacific Ocean. A cold current, the Cromwell is only about 200 miles (322 kilometers) wide and about 700 feet (213 meters) deep, and flows at about 3.5 knots (nautical miles) per hour.

What is the **fastest ocean current**?

The fastest ocean current known is the warm Florida current, part of the Gulf Stream system. It flows from the southern tip of Florida to the north, traveling past Miami, and onward toward Cape Hatteras, North Carolina. It can often reach speeds of 2 to 4 knots, covering 3.3 to 6.6 feet (1 to 2 meters) per second. Some estimates gauge it at 5 knots. Another contender for the fastest current is the Kuro Siwo in the Pacific Ocean.

What are the **slowest currents** in the Earth's oceans?

The slowest currents in the Earth's oceans are those of the deep ocean. The average speed of these currents is about 1 knot (nautical mile), covering 1.6 feet (0.5 meters) per second.

OTHER OCEAN PATTERNS

What is an **upwelling**?

An upwelling is a push of water caused by water temperature differences. Subsurface water, which is colder and denser, wells up to reach the surface—most often replacing wind-displaced surface water. Upwellings also form where currents diverge (split apart) and in places where water is blown away from the coast by strong winds. Upwellings occur most often off the western coasts of continents and in the Antarctic Ocean. For example, the cold Peru (formerly called the Humboldt) Current flows north along the western coast of South America. This causes an upwelling of water off the coast of Peru. Another area of upwelling is the California coast near San Francisco.

What constitutes **nutrient-rich water**?

When water upwells, it often drags along with it nutrient-rich water, or water that contains chemical substances necessary for the maintenance of life in the ocean. The nutrients include materials necessary for the growth of plants, such as nitrites, nitrates, phosphates, ammonia, and silicates. Nutrient-rich water, especially in the cold upwelling waters, often has enough chemical substances to allow plankton to bloom in great numbers at the surface. This in turn feeds the fish and bird population, and creates a fertile breeding ground for all sorts of animals that live in the upwelling area. Consequently, upwellings are prime areas for fishing.

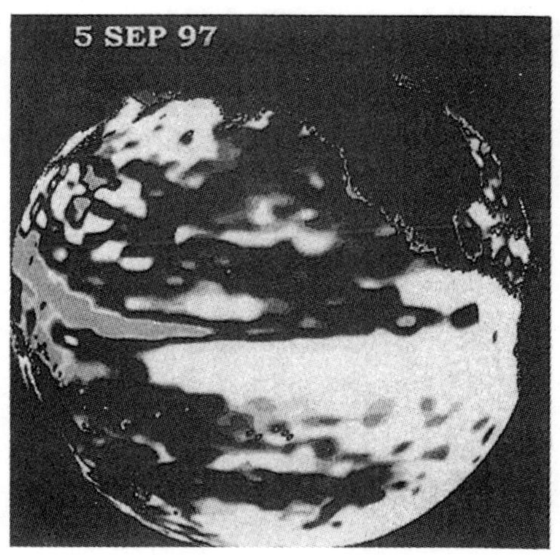

This false-color image of El Niño was taken by the U.S./French TOPEX/Poseidon satellite in mid-September 1997. It shows (in white) the heated surface waters near the equator—a mass that at this time was one-and-a-half times the size of the continental United States. (Also *see* the color insert, where the warmed waters appear in red and white.) *AP Photo/Jet Propulsion Laboratory*

What is **El Niño**?

The phenomenon of El Niño is the unusual warming of a mass of tropical Pacific Ocean surface waters just above the equator and off the western coasts of Peru and Ecuador, South America. El Niño is a Spanish term meaning "the Christ child," or "the little boy," so named because the event seems to peak around Christmastime. The temperature of the ocean waters only rises by a few degrees more than usual, but the increase nevertheless causes major climate changes in pockets around the world. The reason for the phenomenon is unclear. Scientists believe some natural events may be exacerbated by the atmospheric pollutants emitted during larger volcanic events, such as the eruption of Mount Pinatubo (in the Philippines) in June 1991.

It was once thought that an El Niño appears every 15 years—but now it is known to occur every 3.5 to 7 years, and typically lasts 12 to 18 months. Over the past 50 years, El Niño events have begun in 1952, 1958, 1964, 1966, 1969, 1972, 1976, 1982, 1987, 1991, 1994, and 1997. In more recent years, scientists have been able to predict the occurrence of El Niños up to one year in advance. This is possible because of global climate models that were developed to understand the events. But no one yet has been able to predict the intensity of an El Niño.

How do **El Niños impact** the world weather?

Occurring off the western coast of South America, El Niño was once thought to be a localized phenomenon. But it is now understood that El Niño events have a great effect on weather around the world. The following list describes several El Niño events and their effects.

1972: As the waters warmed in the Pacific Ocean, the anchovy population declined, wreaking havoc on the local fishing industry.

1976–77: During this event, severe cold struck the central and eastern United States, and a major drought struck California.

1982–83: This El Niño event caused drought in Australia—the worst one in 200 years. It also worsened a drought that was already underway in Africa. It sent torrential rains to Peru and Ecuador. The United States Pacific coast was inundated with winter storms, while the first typhoon in 75 years hit French Polynesia.

1991–95: This long El Niño changed rainfall patterns around the world. In California, an entire year's precipitation fell in January 1995. The worst floods in a century occurred in Germany, Holland, Belgium, and France. Major droughts occurred in Australia, Indonesia, and southern Africa.

1997–98: This El Niño was classified as a Type 1 (the strongest), involving a high sea-surface temperature. It was also huge—extending from about 160 degrees east to 80 degrees west longitude. What made this El Niño different than the eight other Type 1 events that occurred between 1949 and 1993 was its rapid maturity and growth.

What was the **longest El Niño** ever recorded?

The longest El Niño so far began in 1991 and lasted into 1995, a time when worldwide weather disasters increased. The long-term event may have been helped along by the June 1991 eruption of Mount Pinatubo in the Philippines. Scientists estimate that such long El Niño events occur only every several thousand years—but we don't know enough about past El Niños to know for certain.

What was the **strongest El Niño** every recorded?

The strongest El Niño occurred in 1997 and lasted into 1998; the area of warm surface waters off the coast of South America was also the largest on record. The ocean temperatures averaged almost 11° F (5° C) warmer than usual: temperatures were about 8° F above normal off the South American coast, and 5° F above normal off the coast of Baja, California.

What **problems** can a **strong El Niño** cause?

It is difficult to name all the problems that have occurred during strong El Niño years. For example, in 1982 to 1983, El Niño caused tens of thousands of animals to die off the Galapagos Islands; the droughts in Australia and northern Brazil also caused hardships in those areas. The 1997 to 1998 El Niño probably caused many hurricanes to reach the highest category of hurricane intensity (category 5)—and to move farther north than usual.

What theories are there about **El Niño** events and the end of the **Ice Age**?

Scientists have found that there may be an amazing connection between El Niño occurrences at the end of the Ice Age and glaciers in the Great Lakes. It is known that Lake Huron was once covered by thick ice sheets. When the Ice Ages ended, the glaciers melted and sent water surging through all the Great Lakes; it may be that El Niños contributed to that melting. Evidence from Lake Huron glacial sediments shows that there were several episodes of rapid warming in 10- to 20-year intervals. Scientists believe this is evidence of two events we know about today: El Niño Southern Oscillation (ENSO) and something called the Quasi-Biennial Oscillation (QBO), which alters wind speeds in the upper levels of the atmosphere (the stratosphere) every 2 to 2.5 years.

What is **El Niño Southern Oscillation**?

El Niño Southern Oscillation (ENSO) is a term describing the back-and-forth movement between periodic warming (El Niño) and periodic cooling (called La Niña) in the Pacific Ocean. In other words, the oscillation accounts for the repeated occurrence of both events.

What is **La Niña**?

La Niña is the opposite of El Niño, a time when the surface water of the Pacific Ocean off the coast of South America becomes cooler; in addition, the Earth's trade winds are stronger than normal. Similar to an El

Niño, La Niña changes global weather patterns. Most La Niña years lead to a weaker winter jet stream over the central and eastern Pacific, with fewer storms along coastal California. La Niña is also associated with less moisture in the air, which usually causes less rain along the west coasts of North and South America. But over Southeast Asia and Africa, the usual summer monsoons intensify, and Australia is usually deluged with rains as a result of La Niña.

How do scientists tell whether there is an **El Niño** or **La Niña** in the Pacific Ocean?

Thanks to the launch of special satellites within the past few decades, scientists now have a way to determine El Niño and La Niña events—also detecting if one is getting weaker and the other one stronger. For example, sea surface height readings, which are an indicator of the heat content of the ocean, have been taken by the *TOPEX/Poseidon* satellite. From 1997 to 1998, the rapid cooling occurring in the central tropical Pacific Ocean (indicative of a transition from an El Niño to a possible La Niña) slowed, with the area of lower sea level (or colder water) of La Niña decreasing in size and strength. Thus, scientists could not tell whether or not the La Niña would occur—but the satellite continued to send data to Earth and scientists continued to keep watch. In 1999 most media reported that a La Niña event, albeit a relatively weak one, had unfolded in the Pacific Ocean. But some scientists disputed this conclusion.

What are **"roving blobs"**?

Scientists may talk about "roving blobs" in the Atlantic Ocean; these are massive temperature anomalies, which are thought to be similar to the well-known El Niño and La Niña in the Pacific Ocean. But in this case, these giant patches of warm or cold water drift much more slowly around the North Atlantic—and may affect European weather.

The "roving blob" phenomenon, first described in 1996, was found by studying sea surface temperature records from 1948 to 1992. The "blobs" measure hundreds—or maybe thousands—of miles across and typically last 3 to 10 years; to compare, an El Niño warming persists for 12 to 18 months. These temperature anomalies appear to drift along the path of prevailing ocean currents, but they move at only one-third to one-half the speed of the actual currents. In addition, some scientists believe these "blobs" are also drifting around other ocean basins.

Scientists have traced some of these patches. For example, a cold anomaly developed off the coast of Florida in 1968. It drifted eastward; by 1971, it elongated and hit the coast of Africa. It then traveled south, then west across the tropical Atlantic, reaching the coast of South America in 1975—and dying out within the next two years.

Scientists have yet to determine the cause or details of these massive patches, or if they truly affect climate and weather. For example, some scientists believe a warm patch in the Atlantic Ocean in the late 1950s helped to prolong a Scandinavian drought—but they still need to conduct more research on the phenomenon to determine its effects.

THE OCEAN FLOOR

What is the **ocean floor**?

The ocean floor is the very bottom of an ocean basin. The ocean floor is a multi-level structure that "floats" on top of the denser and more viscous mantle (the part of the Earth that lies below the crust and above the core and which is comprised mostly of unconsolidated material): The layer resting directly on top of the mantle is the oceanic crust, an approximately 3-mile- (5-kilometer-) thick layer of volcanic rock known as basalt; this is the youngest rock on the planet. The next layer, on top of the oceanic crust, is made of lava, and based on its composition, scientists divide it into two different areas: The bottom area, directly on top of the crust, is about 1.2 miles (2 kilometers) thick, and has feeder dikes or lava sheets intruded into other rock; the top area is approximately 0.3 miles (0.5 km) thick, and is made of pillow lava. On top of all of this is a layer of unconsolidated (loose) sediment that was transported there from dry land. This uppermost layer is approximately 1.2 miles (2 km) thick.

What is the **composition** of **oceanic crust**?

The oceanic crust is composed of basalt, a dark gray to black igneous rock that is created by volcanic activity. Basalt can be found at the mid-ocean ridges, where new crust is slowly being created: At the mid-ocean ridges, or spreading centers, the Earth's plates are spreading apart, creating large upwellings of lava, which accumulates to form the ridges. Basalt has a microcrystalline structure, the result of its rapid cooling. It

Are there basins on land?

Yes, there are basins on land. But land basins are much smaller than ocean basins and, of course, they are surrounded by land, with rivers flowing directly into them. Inland, also called interior, basins are usually salty: Rivers pour dissolved salts into the basin and since there is no outflowing river to drain the basin, salts accumulate. For example, the Great Salt Lake in Utah is the low point of the Great Basin, and its water is four times as salty as seawater. The Dead Sea, between Jordan and Israel, is the low point of an interior basin and it is the saltiest lake in the world. Its water is seven times saltier than seawater.

is composed of approximately 50 percent silicon dioxide (SiO_2), 16 percent aluminum oxide (Al_2O_3), and 9 percent calcium oxide (CaO), with many minerals making up the remaining 25 percent.

Where is the **ocean floor**?

Many people think of the ocean floor as all the land beneath the water's surface—including the continental shelf, slope, and rise. However, the true definition of the ocean floor is the land that makes up the ocean basins (in other words, the deep-ocean floor); it is bordered by, but does not include, the continental margins (shelves, slopes, and rises). For example, the North Atlantic Ocean floor does *not* extend from the European and African shorelines to the North American shoreline; instead, it extends, in an east-west direction, from the edge of the European and African continental margins to the edge of the North American continental margin.

What are **ocean basins**?

Ocean basins are the roughly circular or oval-shaped areas of the ocean floor that do not include the continental margins or the structures of the mid-ocean ridges. In other words, the basins stretch from the outer margins of the continents to the mid-ocean ridges. Some are large, such as the Pacific Ocean basin, and measure around 2 to 3 miles (3 to 5 kilometers) deep. Others are smaller, such as the Natal Basin, east of South Africa and south of the island of Madagascar, in the Indian Ocean.

The term "basin" is also used to describe the entire depression that holds the global oceans or it may be used to describe smaller areas of the ocean floor that are surrounded by higher terrain—such as the Gulf of Mexico basin.

Are there any **inland basins below sea level**?

Yes, there are many inland basins below sea level. For example, Death Valley in California is 282 feet (86 meters) below sea level. The Dead Sea is the deepest interior basin, and reaches 1,339 feet (408 meters) below sea level (as of 1996). Because Israel and Jordan gather water from the Jordan River before it flows into the Dead Sea, the water level in this basin drops about 1.5 feet (0.5 meters) annually.

Why don't the **ocean basins fill with sediment** from the continents?

It does seem that over time the ocean basins should fill with sediments from the continents. After all, rivers carry a great deal of sediment from the landmasses to the oceans. But the oceans differ, each accumulating deposits at its own rate. Plus, the accumulation is very slow—not only because the oceans are far from many river sources, but because of the tremendous size of the oceans, it takes a very long time for sediments to accumulate appreciably.

How did the **ocean basins form**?

Scientists believe that the process of plate tectonics (the slow movement of the several plates that comprise the Earth's surface) was responsible for the formation of the ocean basins. The continents are granitic and "float" on a denser mantle rock of silica/magnesium. As a mid-ocean ridge opened, the heavier material from the mantle rose and became seafloor. This material eventually "dove" under another plate at a specific place called a subduction zone (an area where one continental crustal plate moves underneath another); when the mantle remelted (sinking low enough to become molten), it released the lighter rock at the surface as volcanic islands—and eventually formed continents. Therefore, ocean basins—or the places between the continents—have extremely thin crusts when compared with the continental plates; for this reason, they are sometimes referred to as "crustless," or as the "naked plates."

What are some **major features** on the **ocean floor**?

Many people believe, erroneously, that the ocean floor is relatively flat and featureless. The truth is, the land beneath the waves is more diverse than on the continents—in other words, more varied than the land formations we see around us. The ocean floor has ridges, rises, chasms (extending deeper than Mount Everest is high), rolling hills, flat (abyssal) plains, fracture zones, and volcanic islands—and these are just some of the geologic features found under the water.

What caused the **ocean floor's geological features**?

The mountains, trenches, and other features found on the ocean floor were the result of the creation, movement, and loss of the planet's crust. Over billions of years, this ongoing process, which is the result of plate tectonics (the slow movement of the several plates that comprise the Earth's surface), has gradually shaped and changed the face of the Earth—and the ocean floor.

CONTINENTAL DRIFT AND PLATE TECTONICS

When did scientists first realize the **Earth's crust moves**?

In 1620, English philosopher, statesman, and scientist Sir Francis Bacon (1561–1626) noted the similarity between the west coast of Africa and the east coast of South America; he was the first to do so. But his ideas were ignored until the late 1800s and early 1900s. Before that time, the scientific community's prevailing view was that the Earth's surface was static, having been in the same place since the crust cooled.

In 1858, French geographer Antonio Snider-Pelligrini (also seen spelled Pellegrini) first proposed the modern continents had fit together like a gigantic jigsaw puzzle. In his work "Creation and Its Mysteries Revealed," he showed the most obvious fit—the corresponding shape between the west coast of Africa and the east coast of South America. He also believed that the formerly linked continents were split by the Noachian flood (the great flood or deluge mentioned in the Old Testament of the Bible, in the story of Noah's Ark).

The theory that today's continents were once connected was brought up again over the years, but it was ignored by most scientists. In 1912 Alfred Wegener (1880–1930), a German meteorologist and geologist, published a book titled *The Origins of Continents and Oceans.* The book expanded on the theory put forth by Snider-Pelligrini and others, proposing that all the continents had once fit together in a large supercontinent, which Wegener called Pangea (or Pangaea, from the Greek *pangaia,* meaning "all Earth"). He believed Pangea had broken up into smaller landmasses, with the shape and location of today's continents a result of their movement across the Earth's surface. He called this process continental drift—a term that is believed to have been inspired by the way ice floes move across the surface of the water. However, Wegener could not explain the mechanisms driving this movement and his theory was not taken seriously. By the 1950s and 1960s, scientists finally accepted Wegener's general theory, since by this time enough evidence had accumulated to support the theory and scientists had worked out a mechanism, called plate tectonics, to explain the movement of the continental plates.

What are **continental drift** and **plate tectonics**?

The most accepted theory of continental movement is called continental drift; the theory of its mechanism is called plate tectonics. The term tectonic comes from the Greek word *tekton,* meaning "builder," a reference to the building of mountains from the movement of continental plates.

The continental plates move laterally across the face of the planet. Boundaries between plates include the mid-ocean ridges (or spreading centers), subduction zones (areas where one continental crustal plate moves underneath another), and areas in which plates slip by—or pass—one another.

The theory of continental drift was first suggested by German geophysicist Alfred Wegener (1880–1930). The evidence for this movement has been collected by observation satellites, which allow scientists to use sophisticated laser-ranging instruments to track minute movements in the plates that make up the Earth's crust. These plates move just fractions of an inch to inches each year.

No one can truly explain the reason (or reasons) why the continental plates, and thus the continents, move (or "drift") across the face of the Earth. But scientists are certain that the plates do move—which has prompted numerous theories to explain why they move. One theory suggests that the continents are situated on large tabular blocks of the Earth's crust. The mechanism that drives the movement of these blocks is the fluid mantle (the part of the Earth that lies below the crust and above the core, and is comprised mostly of unconsolidated materials) underneath. The interactions between the plates, which occurs at or near their boundaries, give rise to many of the features found on land and the ocean floor.

What evidence led to the **acceptance** of **continental drift** and **plate tectonics**?

Evidence accumulated over time and by the 1960s there was a general acceptance of the theories of continental drift and plate tectonics. Satellite data, fossil findings, and seafloor spreading indicated to the scientific community that the Earth's plates had moved and continue to do so. The theory was strengthened by further evidence—the discovery of magnetic anomalies in ocean rocks.

And although no one can, at this stage, definitively explain the reason (or reasons) why continental plates—and thus the continents—move across the face of the Earth, scientists are certain the plates do move.

How did fossils help prove the theory of plate tectonics?

Similar fossils were found on widely separated continents, indicating that separate species either developed identically across the far-flung continents (a notion that is highly unlikely), or the continents were in contact millions of years ago when these species developed. For example, fossils found in South America were found to be related to those in Australia and Antarctica; others discovered in Antarctica were related to those in Australia. These findings seemed to indicate that the landmasses had once been joined together: Species roamed freely across the surface, died, were buried by sediment, and became fossilized. As the continents drifted apart, they carried the organisms' remains—leading to the finding of widely separated, but identical, fossils.

How do **satellites** help **prove the theory of plate tectonics**?

Observation satellites, which use sophisticated laser-ranging instruments, allow scientists to track minute (extremely small) movements in plates. Without the data sent back to Earth by these satellites, the almost infinitesimally small movements of the plates could not be detected.

What are **active** and **passive margins**?

Active and passive margins describe the different characteristics of continental margins that are a result of plate tectonics (the slow movement of the plates). For example, the continental margin (or edge) of the Atlantic Ocean is a passive margin, with a well-developed continental shelf, slope, and rise. This type of margin developed as continents split and drifted apart, and an ocean basin was created between them. A passive margin is the "trailing edge" of a plate; these margins are also called "constructive margins" because sediment deposited from the landmass builds the margin outward—toward the sea.

An example of an active margin can be found on the western (Pacific) side of the South American plate. Here, the continental shelves are narrow, the slopes are deep and narrow, and the rises are frequently missing. These features are characteristic of colliding plates. Active margins are areas of frequent geological activity, with numerous volcanoes, earthquakes, and mountains. An active margin, in other words, is the "leading edge" of a plate.

What is **plume tectonics**?

Plume tectonics is how scientists describe the behavior of the magma (the hot, liquid rock) in the Earth's mantle—the below-the-surface activity that causes the plates to move. They use seismic tomography, a technology that scans the Earth's interior with seismic waves, much like a CT scan examines the human body, to track the movement in the magma. Based on this and other studies, mantle convection (heating) occurs in association with "hot plumes" and "super-plumes" that move up from the boundary between the core and mantle, and "cold plumes" that drop toward the boundary. For example, beneath Africa and Tahiti, there are hot super-plumes rising, creating volcanoes and moving plates; beneath Japan, a cold plume falls, creating a subduction zone (an area where one plate moves beneath another).

What are the **major plates** of the **Earth's crust**?

The Earth's crust is divided into 13 major plates: the North American, Eurasian, African, South American, Pacific, Nazca, Cocos, Somali, Australian, Bering, Philippine, Arabian, and Antarctic. Each plate boundary has a specific direction of movement. There are also other, smaller plates within the larger plates.

What **happens** at the **plate boundaries**?

There are three main events that occur at the boundaries of moving plates: 1) The plates may separate or pull apart; 2) they may collide with other plates, with one plate subducting (moving underneath) the other—usually forming deep trenches or building mountains in the process; or 3) two plates can slip past one another, moving two pieces of land in opposite directions.

What happens when **two of the Earth's plates separate**?

At the boundary where two plates separate, molten rock from the mantle (the part of the Earth that lies below the crust and above the core) rises from deep within the Earth. As it flows out, it accumulates to create a mid-ocean ridge; and as the plates continue to separate, new oceanic crust is continually created. One well-known example of this geological process is the Mid-Atlantic Ridge, a long chain of volcanic mountains cutting through the center of the Atlantic Ocean. On land, the boundaries where plates are separating are called rift valleys. Eastern Africa's Great Rift Valley is one example.

What happens when **two of the Earth's plates collide**?

At the boundary between two colliding plates, one plate can be forced under the other—causing the oceanic crust to sink into the mantle (the part of the Earth that lies below the crust and above the core). This process is called subduction, and the area where it occurs is called a subduction zone. These areas generate many earthquakes and volcanoes, and form deep-ocean trenches. One well-known subduction zone lies along the west coast of South America. This is where the Nazca plate is subducting (moving underneath) the South American plate, creating a subduction zone. This activity once created the volcanic Andes Mountains and still creates frequent earthquakes.

What happens when **crustal plates slip** by one another?

At some boundaries, the plates are neither separating nor colliding—sometimes they move by each other (side by side), in opposite directions, in a process called slip. When these moving plates get stuck, a large amount of energy builds up along the boundary; when they let go or move slightly, earthquakes result. The most famous example of a boundary between two slipping plates is the San Andreas Fault, which runs through California; here a part of the North American plate slides by the Pacific plate.

How was the process of **seafloor spreading** discovered?

In the mid-twentieth century, American educator and geologist Harry Hess (1906–69) took part in an ocean-drilling project called MOHOLE. Examining material from the ocean's bottom layers, Hess discovered that ocean floor rocks are younger than those found on the continental landmasses. He also found that the ages of the ocean floor rocks vary—the farther from the mid-ocean ridge, the older the rocks; and conversely, the youngest rocks are found near mid-ocean ridges.

In a paper titled "Further Comments on the History of Ocean Basins," which Hess delivered to the Geological Society of America in 1963, he theorized that the seafloor was spreading. He suggested that seafloor spreading was caused by magma (hot, liquid rock) erupting from the Earth's interior along mid-ocean ridges. This newly-created seafloor slowly spreads away from the ridges; it later sinks back into the Earth's interior at deep-sea trenches.

How did **magnetic anomalies** support the **theory of seafloor spreading**?

The theory of seafloor spreading was later confirmed by the measurement of magnetic anomalies in ocean floor rocks. When molten lava is expelled from an ocean ridge, it creates new ocean floor. The magma (the hot, liquid rock) then cools and elements within the rock line up with the planet's prevailing magnetic field, retaining the orientation present at that time. Since the planet's magnetic field changes over time, evidence of this would show up in the ocean floor rocks.

In rocks from the ocean floor, scientists discovered a pattern of stripes representing these magnetic anomalies—providing a record of the changes in the orientation of the magnetic field over time. These patterns spread out symmetrically on either side of the Mid-Atlantic Ridge, a long chain of volcanic mountains cutting through the center of the Atlantic Ocean. Scientists concluded that the pattern and distribution of these rock stripes, showing the magnetic field reversals over time, could only have occurred if the seafloor had been spreading apart over millions of years.

How many times have the **Earth's magnetic poles reversed**?

The Earth has reversed its north and south poles about 177 times over the past 85 million years. Based on the study of certain rocks and their magnetic properties, scientists estimate that a magnetic reversal took place only 2 million years ago. No one knows how or why the magnetic fields switch places—and we definitely don't know what happens when the fields *do* change.

Where does the **crust** from spreading seafloor eventually **end up**?

The crust created by the spreading seafloor eventually subducts (moves beneath) a continental plate; from there, the crust sinks to the Earth's mantle, where it again becomes molten. Over time, this molten rock may be "recycled" into new seafloor at mid-ocean ridges—like a giant conveyor belt working in the Earth.

Are there any **other explanations** for the **movement of the continents**?

Yes, there are other explanations for the movement of the continents—because not everyone agrees with the mechanism of plate tectonics. One reason is that, although the idea of moving plates seems sound, the mechanisms behind plate tectonics are virtually unknown. Thus some scientists believe in continental drift, but not in plate tectonics.

Another possible explanation for continental drift is that the Earth is expanding. This would give the illusion that the continents move across the surface of the planet (although no one can explain why or how the Earth is expanding). Another theory is "surge tectonics," in which there is a sudden surge of plate movement, as opposed to a constant flow. And still another suggestion is that there is no continental drift—that the continents have *always* been in the same positions they are in now.

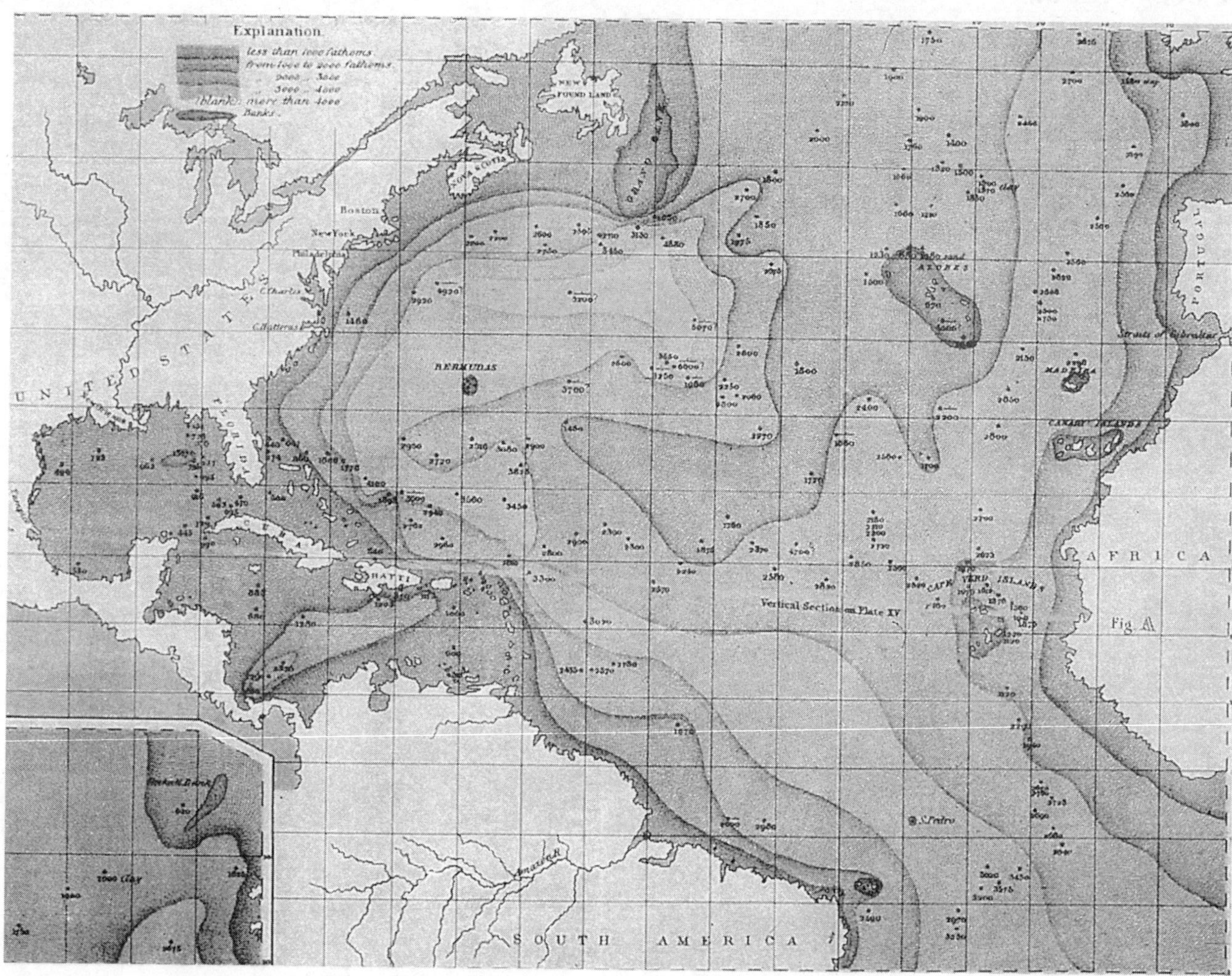

American naval officer Matthew Maury's 1855 bathymetric map of the Atlantic, the first of its kind, gave birth to the idea of the Telegraphic Plateau—a submarine land formation from Canada to the British Isles, across which the first transatlantic cable was laid. ***NOAA Central Library***

RIDGES AND RIFTS

What is a **mid-ocean ridge**?

A mid-ocean ridge is essentially an underwater volcanic mountain range, found at the boundaries between separating plates. These ridges girdle the globe, like stitches on a baseball, and are usually sharply crested with steep sides. All together, they form the longest series of mountains on Earth, extending more than 35,000 miles (56,000 kilometers) in length.

What is the **difference** between **mid-ocean ridges** and **rises**?

Both mid-ocean ridges and rises are used to describe underwater mountain chains. But physically, rises tend to have a gentler topography and lack the central rift valleys found in ridges. For example, the Mid-Atlantic Ridge has a deep rift valley along its crest, measuring about 6.2 miles (10 kilometers) wide, and with walls about 1.9 miles (3 kilometers) high. Such dramatic topography would not be found on a rise.

What are some of the **major mid-ocean ridges** and **rises**?

Some of the major mid-ocean ridges and rises run in a north-to-south direction. These include the Mid-Atlantic Ridge in the Atlantic Ocean, the Reykjanes in the North Atlantic and Arctic oceans, the East Pacific Rise in the Pacific Ocean, and the Mid-Indian Ocean Rises in the Indian Ocean. Others include the Carlsberg Ridge, which runs southeast from the Arabian Plate to the Mid-Indian Rise, and the Southeast Indian Ridge, located between Antarctica and the Indian and Australian plates.

When were **mid-ocean ridges discovered**?

The mid-ocean ridges remained unknown to humans until the nineteenth century. In the 1850s, American naval officer and oceanographer Matthew Fontaine Maury (1806–73) became the first to show that the ocean bottom isn't smooth and featureless. He discovered and named the Dolphin Rise in the Atlantic, the first indication of the Mid-Atlantic Ridge, which cuts through the Atlantic Ocean from north to south (extending from roughly 55 degrees north latitude to 55 degrees south latitude).

But it wasn't until the voyage of the H.M.S. *Challenger* (from 1872 to 1876) that the single, continuous nature of the submarine mountain range, now known as the Mid-Atlantic Ridge, was discovered. And it took until the 1960s and 1970s before the entire ridge was mapped and studied.

What are some **common features** of the **mid-ocean ridges**?

The underwater mountain ranges that are the mid-ocean ridges have a unique characteristic: While continental mountain ranges normally have a single, prominent line of peaks, underwater ranges associated with mid-ocean ridges have two lines of peaks, separated by a rift valley that can range in width from 15 to 30 miles (24 to 48 kilometers). Some of the mid-ocean ridges are longer than any continental mountain range. If these ridges were strung together, end to end, they would form an undersea mountain range that would extend some 35,000 miles (56,000 kilometers) in length. The individual ridges' widths vary from 500 to 1,500 miles (805 to 2,414 kilometers). The peaks of most mid-ocean ridges are approximately 8,000 feet (2,438 meters) below the ocean's surface. Some can rise as high as 12,000 feet (3,658 meters) from the ocean floor, with their peaks ascending above the water level to form islands. Iceland and the Azores in the Atlantic and the Galapagos Islands in the Pacific are actually the peaks of underwater mountains.

How do **mid-ocean ridges form**?

According to the theory of plate tectonics (the movement of the Earth's plates), mid-ocean ridges form when two continental plates pull apart, and magma (hot, liquid rock) wells up from deep in the Earth, creating new crust—and thus, new seafloor. This moving magma also pushes the plates apart (in what is called seafloor spreading), contributing to the movement of the continental plates; is responsible for the continuous build-up of the ocean's crustal plates; and in certain areas, creates a buildup of lava, which eventually forms a mid-ocean ridge. Although the number of oceanic volcanoes responsible for this upwelling of molten rock is unknown, it is thought to be quite high.

What is one of the **most active** areas along the **Mid-Atlantic Ridge**?

One of the most active areas along the Mid-Atlantic Ridge, which extends from roughly 55 degrees north latitude to roughly 55 degrees south latitude, is near the island of Iceland.

What is a **graben**?

A graben is a low block of rock that is bounded on two sides by parallel faults that create tall, cliff-like scarps. These fault scarps—which are slopes that occur at the boundary between two plates or at fractures in the Earth's crust—are caused by the movement of the plates.

What is a **rift valley**?

A rift valley is an area where plates are separating; it is actually a graben—with fault scarps (slopes) on either side. Rift valleys are created by magma (hot, liquid rock) that wells up from the Earth's mantle and hardens. They can occur in the ocean, along a mid-ocean ridge (such as the Mid-Atlantic Ridge) or on land, where there are two main areas of this geological activity: around Iceland, which is a peak of the Mid-Atlantic Ridge that has grown tall enough to appear above the surface of the water; and in eastern Africa's Great Rift Valley, which stretches more than 3,000 miles (4,800 kilometers) from Syria to Mozambique.

Is a **new ocean forming**?

Scientists believe that the Great Rift Valley in eastern Africa is where the next ocean will form. As the sides of the valley move farther apart, the floor (or graben) between them will lower. Currently, the Red Sea fills part of the Great Rift, and as the rest of the valley sinks, scientists predict that water will pour in. This will flood the valley, eventually splitting Africa in two—and creating a new ocean.

UNDERWATER VOLCANOES AND EARTHQUAKES

What is a **volcano**?

A volcano is a vent in the Earth's surface; it can occur either on land or under water. At these openings, eruptions occur—caused by pressures beneath the Earth's surface. Gases, ash, dust, and magma (hot, liquid rock from deep within the planet) are forced out, creating numerous geological features—the most prominent and spectacular being the volcano itself, which is an accumulation of the debris that erupted from the vent.

The word *volcano* comes from the Vulcano Island in the Tyrrhenian Sea, off the northern tip of Sicily (or off the "toe" of Italy's boot). Here there is a volcano that, though not active now (the last time it erupted was 1890), was active in ancient times, causing Romans to believe it was the entrance to the nether regions and the domain of Vulcan, the god of fire, who fashioned armor for the other gods.

What is a **high island**?

The term "high island" describes an island of volcanic origin in the Pacific Ocean. High islands can be active or dormant. The Hawaiian, Bougainville, and Solomon islands are all high islands.

How do **volcanoes form**?

Volcanoes can form at the boundaries of plates or in the middle of plates. At a boundary formed by separating plates, volcanoes can be created when molten rock flows to the surface through the openings. The numerous volcanic mountains associated with mid-ocean ridges were formed in this way. Where plates collide, one plate may subduct (move beneath) the other, and as the subducted plate sinks into the Earth's mantle below, it becomes molten and the resulting magma (hot, liquid rock) rises up to form volcanoes. Volcanoes that were formed in this way are normally found along the edges of continents, such as in the Andes

Mountains of South America, or rising from the ocean floor, such as the islands of Japan. Volcanoes also form in the middle of plates due to the presence of "hot spots," creating islands or chains of islands, such as the Hawaiian Islands.

The actual mechanism behind magma formation and eruption is still being debated by scientists, as are ways of predicting volcanic eruptions. Volcanoes can cause widespread destruction and loss of life, but they are also one of the few ways we have to study the interior of our planet. Our current drilling techniques can only penetrate approximately 6 miles (10 meters) into the crust, while the magma of the Hawaiian volcano Kilauea originates at depths of at least tens of miles (or at least 20 kilometers).

Why are **volcanoes important features** of the ocean floor?

Volcanoes are important features of the ocean floor because they are newly created land, continually rising up out of the deep bottom. Whether their summits remain below the surface or rise above the waves, they provide the foundation for unique ecosystems that thrive on and around them. The active volcanoes continue to add energy, mostly in the form of heat, to the environment around them; this energy is used by many forms of life. More than 80 percent of the Earth's volcanic activity occurs in the deep-ocean environment.

Are there different **types of volcanoes**?

Yes, there are numerous types of volcanoes, which are mainly categorized by their physical features. Some common examples of volcanoes include shield (low-sloping), caldera (having a basin or crater at the summit), fissure (having vents in the side), lava dome (resembling a raised bubble), composite (also called a stratovolcano, which has alternating layers of ash and lava), and ash-cinder (a straight-sided ash cone). Volcanoes can also be a combination of several types—such as a composite caldera. Certain types, including caldera, composite, and ash-cinder, usually experience violent eruptions (spewing pumice and ash), while the other types have less explosive eruptions, producing lava flows that cover large areas.

This black lava sand beach in Hawaii was later destroyed by lava flows. *NOAA/Commander John Bortniak, NOAA Corps (ret.)*

Where are some of the **largest calderas** on Earth?

The ash-flow calderas (great basins at the summits of volcanoes made mostly of ash) are the largest calderas on the planet. They are characterized by their very low relief, broadness, and immense ash deposits. The volcanic calderas of Yellowstone National Park (one of the largest volcanic fields on Earth) are examples of this type of volcano.

What **types of lava** come from **above-water** volcanoes?

There are two basic types of lava, which is ground-flowing magma (hot, liquid rock), that flows on land. Their Hawaiian names are *pahoehoe* (pronounced pa-hoy-hoy) and *aa* (ah-ah). Pahoehoe is a smooth lava that forms a ropy surface (the lava cools to form wrinkles); it comes from higher-temperature and lower-volume eruptions, and has low viscosity, allowing it to flow easily and form a skin. It usually moves at approximately a yard (or meter) per minute. But, under the right conditions, such as on a steep slope, it can move at speeds up to 14 miles (23

Hawaii's lava flows encroaching on the ocean. *CORBIS/Morton Beebe*

kilometers) per hour. The thickness of a pahoehoe deposit is typically about 12 inches (31 centimeters).

Aa lava is sharp, rough, and coarse, and it normally occurs under the opposite conditions of pahoehoe. Aa tends to flow in surges, with the front advancing slowly, building up its height as it moves along at a rate of a few yards (several meters) per hour. Then, suddenly, it will move quickly forward, returning to its original thickness while covering ground at the rate of a hundred yards (meters) in a few minutes. The normal thickness of an aa flow is 6.5 to 16.5 feet (2 to 5 meters); flows are usually large, extending more than 100 yards (300 feet, or 91 meters) in width, and are often fed by smaller streams of lava.

What is **pumice**?

Lava pouring out from a volcano can eventually flow into the ocean and cool very rapidly. This creates a volcanic glass called pumice, a very light rock often found floating around volcanic islands. The rock has the chemical composition of a granite (quartz and feldspars), but it has a low density that allows it to float. Because the lava cools so fast, it is packed with bubbles; as it solidifies, the bubble stay, giving the rock its low density, lightness—and buoyancy.

What happens when **lava flows underwater**?

The flow of lava underwater is different than on land. This is because the water temperatures are colder and the pressures in the deep ocean are greater than that (air pressure) on land. Because of these factors, underwater lava develops very specialized features, including pillow lava and submarine lava pillars.

What is **pillow lava** and how does it **form**?

Pillow lava is basalt (black volcanic rock) that has cooled, forming relatively small mounds. It only forms under water. As molten lava erupts from the depths and is deposited into the ocean, the surrounding water cools the exterior of the lava, forming a flexible but glassy crust in the shape of a pillow; the interior remains molten. As the amount of molten material increases in a particular pillow, the pressure increases, expanding the pillow and eventually breaking a hole in its glassy crust. The lava slowly flows out of this opening, the water cools the exterior of the outpouring lava, forming another pillow, and the process is repeated. This sequence continues as long as the eruption goes on, many times producing a thick deposit of pillow lava over a large area. The lava itself can originate from an underwater eruption, such as from a mid-ocean ridge or underwater volcano, or from lava that flows from the land into the ocean.

What are **lava pillars**?

Underwater lava pillars (also called basalt pillars) are gnarled columns of a black volcanic rock called basalt, some up to 50 feet (15 meters) tall.

They range from thin and spindly to as thick and stout as a giant redwood tree; they also form as lava pillar archways. The outside of the pillars are coated with a paper-thin layer of black volcanic glass; the inside is a peppery gray, lined with a network of fine cracks. They are invisible in the blackness of the deep oceans until lit by a light source. They were first discovered by the lights of a passing submersible.

Pillow lava is formed underwater by basalt (black volcanic rock) that has cooled into small mounds such as those pictured here (off the coast of Hawaii). *NOAA/OAR National Undersea Research Program*

These huge formations are created when lava erupts onto the seafloor, but the actual mechanism by which they are created is still under study. One theory is that during a submarine eruption, water trapped below the growing layer of magma (molten rock) squirts through gaps in the lava flow, allowing the pillars to grow. As more lava flows out, the lava layer on the seafloor thickens; it continues to grow upward around the jets of water, leaving the duct open. Because the water is much cooler than the lava, it "freezes" the lava into pipe-like cylinders that grow around the jets—the building of them stops only when the flow of molten rock subsides.

Scientists hope to use pieces of lava pillars gathered by submersibles to better understand mid-ocean spreading centers. The details of these areas are relatively unknown because they are located so deep on the seafloor—at depths of 8,203 to 11,483 feet (2,500 to 3,500 meters).

Are any **volcanoes forming** in the **oceans today**?

Yes, there a number of volcanic eruptions occurring on the floor of the oceans today. In fact, scientists estimate that some 80 percent of our planet's volcanic activity happens in this deep, hidden realm. One of the most spectacular phenomena associated with volcanic activity is the growth of huge, localized volcanoes—some of which become large enough to poke above the waves, forming islands (called seamounts) and sometime island chains. Examples of currently forming volcanoes are the Loihi Seamount, off the coast of Hawaii, and the Axial Seamount, off the north coast of Oregon.

Is there a **volcanic basis** for the **legend of Atlantis**?

Although most scientists agree that the underwater city of Atlantis did not exist, they may have discovered the origin of the myth. In ancient times, Minoans lived on the island of Thira (or Thera)—a Greek island in the Aegean Sea, where the volcano Santorini is situated. Around 1470 B.C.E., the volcano had a huge eruption and all that remains of the island today is a large, crescent-shaped caldera (volcanic crater). Centuries after the eruption, the Greek philosopher Plato (c. 428–348 or 347 B.C.E.) described the island's "disappearance." Over the years, his account may have evolved as the myth of Atlantis—the city that sunk into the sea.

Are there **volcanoes** elsewhere **in the solar system**?

Yes, there are other volcanoes on planets and satellites in the solar system. Venus has many—seen from spacecraft such as the *Magellan,* which took radar images of the planet. Mars also has volcanoes, including one of the largest known in the solar system; it is called Olympus Mons. The *Voyager* 1 and 2 spacecraft to the outer planets discovered more volcanoes. In fact, in the late 1970s, a *Voyager* captured images of an eruption on Io, one of Jupiter's moons! Since that time, ground-based telescopes and the *Galileo* spacecraft around Jupiter have imaged many other such eruptions on the small moon.

What is the **Ring of Fire**?

The Ring of Fire is the name given to the periphery of the Pacific Ocean—a belt of seismic activity; it is also called the Circle of Fire. Counterclockwise, it extends from the southern tip of South America, north to Alaska, then west to Asia, south through Japan, the Philippines, Indonesia, and to New Zealand. (Hawaii is in the middle of the ring.) The Ring of Fire is an area where plates are subducting (moving underneath each other), creating numerous volcanoes—more than half of the Earth's active volcanoes. For example, Japan has 77 subduction-generated volcanoes, while Chile has 75. In addition, Kamchatka has 65, Alaska and the Aleutian Islands have 68, and the western United States has 69 such volcanoes. Some famous volcanoes along this ring are Japan's Mount Fuji; Indonesia's Galunggung; Mount Katmai and Augustine in Alaska; and Mount St. Helens in Washington state.

Which **volcano** is the **most massive**?

The most massive volcanic mountain—indeed, the most massive mountain of any type on Earth—is Mauna Loa (*see* photo next page), located on the Big Island of Hawaii. Created by a hot spot located underneath the ocean, this mountain stands 13,679 feet (4,169 meters) above sea level. It rises a total of approximately 30,000 feet (9,144 meters) from the ocean floor and occupies almost 10,000 cubic miles.

Which **volcanic eruptions** have been the **most destructive**?

Since the nineteenth century, the most destructive, or deadliest, eruptions (listed in order of destruction, from worst to least) occurred at Mount Tambora, Indonesia, on April 5, 1815; Krakatau, Indonesia, on August 26, 1883; Mount Pelee, Martinique, on August 30, 1902. The most recent deadliest eruption occurred on November 13, 1985, in Nevada del Ruiz, Colombia; 23,000 people died.

Are the **death tolls** from **volcanic eruptions on the rise**?

Yes, similar to most deaths from natural disasters, there is a rise in the death toll from volcanic eruptions. This is because as the global population increases, there will be more people living in the direct path of such explosive events. Scientists estimate that between 1600 and 1900, the death toll from volcanic eruptions was about 315 people annually. In the twentieth century alone, the number has increased to about 845 people annually—and the number is expected to keep rising.

The volcanic Mauna Loa, located on the Big Island of Hawaii, is the most massive mountain (of any type) on Earth—its base lies on the seafloor. *NOAA/Commander John Bortniak, NOAA Corps (ret.)*

What are some **mid-ocean volcanic islands**?

There are numerous mid-ocean volcanic islands. One of the largest is Iceland, off the southeast coast of Greenland. This huge island was created by volcanic material from roughly 200 active volcanoes—all from eruptions along the spreading Mid-Atlantic Ridge. The Azores Islands are also situated along the Atlantic Ocean's mid-ocean ridge, just off the coast of Portugal; as are the islands in the Tristan da Cunha group, in the South Atlantic Ocean.

What is **Surtsey**?

Surtsey is a volcanic island just off the shore of Iceland, in the North Atlantic Ocean. The island appeared only recently—on November 16, 1963, to be precise—during a spectacular display of erupting magma and steam in the ocean off Iceland's southern coast. The island took about 4 years to grow, reaching 492 feet (150 meters) above sea level by 1967, with an area of 2 square miles (5 square kilometers). Early in the island's life, ocean waves caused a good deal of erosion, but the core quickly solidified. Surtsey, like Iceland, lies over the Mid-Atlantic ridge; it is now a permanent island, colonized by plants and sea life.

Did any **other islands** grow around **Surtsey**?

Yes, two other smaller islands formed near Surtsey: The Syrtlingur (or "Little Surtsey") in May 1965 and the Jólnir (or "Christmas Island") in December 1965. These islands were not as lucky as Surtsey. Both soon disappeared after the volcanic activity ceased in the area, the new lava worn away by the wave action of the sea. For example, Jólnir disappeared in August of the following year—its life was about 8 months.

Why are **Iceland's volcanoes unique**?

The volcanoes on Iceland are all on the Mid-Atlantic Ridge—but they are not like the common cone-shaped volcanoes found around the world. Icelandic volcanoes are actually fissures hidden beneath a glacier that covers approximately 8 percent of the island. This means eruptions—which have occurred every five years for the past 1,100 years—can melt large parts of the overlying ice, unleashing huge floods.

What is a **hot spot**?

A hot spot is another process created by plate tectonics—the movement of the Earth's plates. Hot spots lead to the creation of volcanoes on the ocean floor—but they are distinct from separating or colliding plates. Canadian geophysicist J. Tuzo Wilson (1908–93), one of the pioneers of the theory of plate tectonics, sought to explain the presence of volcanoes that formed in the middle of a plate instead of along a plate boundary. He postulated the presence of what he called a "hot spot," a localized, stationary upwelling of magma (molten rock) from the Earth's mantle.

Although the mechanism is still debated, many scientists believe a hot spot is created by an upwelling of magma, which flows up through weak areas in the overlying plate. This often leads to the formation of a volcano. Over millions of years, as the plate continues to move slowly over this hot spot, a chain of volcanic islands is born. Eventually, the oldest of these often inactive volcanoes erodes and "sinks" below the water's surface.

Where are **hot spots located**?

Hot spots are located around the world, and some of them are quite well known. Scientists believe the Hawaiian Islands are a chain of volcanic islands created by the presence of a hot spot, which has generated almost 200 volcanoes over a period of 75 million years. The most recent above-surface creation from this hot spot is the Big Island of Hawaii. Other places that may be hot spots include the Galapagos and Society Islands, and Yellowstone National Park in Wyoming.

Where does the **Hawaiian Island hot spot originate**?

Scientists may have recently pinpointed the origin of the Hawaiian hot spot at the boundary between the mantle (the part of the Earth that lies below the crust and above the core) and the metallic core of the planet. This plume of molten material is approximately 1,800 miles (2,900 kilometers) beneath the crust, where the molten outer layer of the core heats the rock at the base of the mantle. Data also suggest that the molten plume of material first flows horizontally toward the base of the Hawaiian hot spot, then rises vertically.

Using a technique called seismic tomography, scientists analyzed this hot spot by studying seismic waves generated by earthquakes in the region of Tonga and Fiji; the waves traveled through the deep mantle beneath the Hawaiian hot spot before reaching recording stations in Oregon and California. They also analyzed the waves for polarization; this showed, for the first time, variations at the base of the mantle—indicating the molten material's localized transition from a horizontal to a vertical flow.

What is the **newest Hawaiian Island**?

Currently, the volcano called the Loihi Seamount is building just off the islands, and will eventually become the newest Hawaiian island. This mountain—about 20 miles (32 kilometers) off the coast of the Big Island of Hawaii—rises some 17,000 feet (5,182 meters) above the ocean floor and is about 3,000 feet (914 meters) below the ocean's surface. In several thousand years, Loihi will poke through the surface of the ocean water to become the newest island in the Hawaiian chain.

Are there any **hot spots** in the **Indian Ocean**?

Yes, another well-known hot spot track extends from India to the island of Réunion, in the Indian Ocean. About 67 million years ago, present-day India was above the hot spot. A great deal of basaltic lava erupted, creating a huge volcanic field called the Deccan Trapps. As the plate moved to the northeast over the hot spot, more volcanic centers formed, creating features including the Maldives (around 57 million years ago), Chagos Ridge (48 million years ago), Mascarene Plateau (40 million years ago), Mauritius Islands (18 to 28 million years ago), and in the last 5 million years, the volcanoes of Pito des Neiges and Piton de la Fournaise, which make up the island of Réunion.

What are **black smokers?**

Black smokers are features of the deep-ocean floor, occurring in areas where there is volcanic activity. They are created by hydrothermal vents: At these vents, super-hot water erupts in a plume that looks like a column of black smoke; but it is actually hot, salty, mineral-laden water—at temperatures upwards of 700° F (370° C)—which is released from the vent and encounters the much cooler seawater. The sulfide and sulfate minerals in the water give the plume a dark color. As these minerals are deposited around the plumes, sulfide chimneys build up, and from these chimneys the black water continually flows.

When did scientists **discover black smokers**?

The first black smokers were discovered in 1977 off the Galapagos Islands, off the coast of Ecuador, in the Pacific Ocean. Scientists were investigating the ocean floor in the submersible *Alvin,* a mini-submarine built by the Woods Hole Oceanographic Institution. Near the Galapagos, they noticed these "smoking chimneys." Since that time, several chimneys have even been recovered from the ocean floor—some measuring 5 feet (1.5 meters) tall, and ranging in weight from 1,200 to more than 4,000 pounds (545 to more than 1,816 kilograms).

This black smoker (the super-heated plume from a hydrothermal vent) was photographed on Endeavor Ridge—an underwater land formation off the California Coast. *NOAA/OAR National Undersea Research Program; P. Rona*

Where are **black smokers found**?

Black smokers tend to occur on sulfide mounds, which, over time, build up along mid-ocean ridges. These mounds—ranging anywhere from about the size of a pool table to the size of a tennis court—grow in fields of hydrothermal vents (a volcanic opening where super-hot water escapes from the depths of the Earth). Although black smokers have only been found on mid-ocean ridges, not every ridge has black smokers. Scientists are still trying to figure out why some ridges have these hydrothermal vents and others don't. In addition, scientists believe only a relatively few black smokers have been discovered so far, since only a small portion of the world's approximately 31,070 miles (50,000 kilometers) of ridges have been explored. More of these features will doubtless be found in future explorations.

Why are **black smokers important**?

Black smokers are important because they form the foundation for a unique ecosystem. This ecosystem is located deep in the darkness of the ocean bottom, in the most extreme environment possible for life.

Scientists studying these chimneys, where super-heated waters escape from hydrothermal vents in the seafloor, discovered unique life forms dwelling on and around the area—life forms such as tube worms and exotic shrimp thriving without sunlight, under immense pressure (from the surrounding waters), and in high water temperatures. Also, scientists found different types of microbes living in the interiors of these chimneys, some of which are heat-loving, or thermophilic; these microbes are some of the most primitive forms of life found on Earth.

The ecosystem around the black smokers is the only one on Earth that does not rely on the rays of the Sun as its foundation. All the organisms living around a hydrothermal vent (a volcanic opening where super-hot water escapes from the depths of the Earth) are dependent on bacteria that use hydrogen sulfide from the hot water plume as their primary energy source.

The circulation process initiated by these hydrothermal vents is also important to our planet: The vents transfer large amounts of heat and chemicals from the interior of the Earth into the oceans, which in turn, greatly influences the properties of the ocean's water—and regional climate.

Is there a connection between **black smokers** and the **origin of life**?

Yes, there may be a connection between black smokers and the origin of life on Earth. Scientists believe the presence of unique organisms around black smokers indicates that life could exist—and evolve—without the Sun's rays. And some scientists have even speculated that life on Earth originated around early black smokers. These areas were "safe havens" for these forms of life, a place to hide while other organisms went extinct due to changes in the climate or other causes. Because of these findings, many scientists now believe life may be present at deep-ocean vents on other planets or satellites in our solar system. For example, such organisms may live around vents on Jupiter's moon Europa, a satellite that may hold an immense ocean underneath its icy surface.

What **unique phenomenon** is associated with **black smokers**?

Recently, scientists have discovered an extremely faint glow of light is given off by black smokers. This light is imperceptible to humans, and can only be recorded using cameras sensitive to low light levels. The cause of the faint glow of light emanating from black smokers is still a mystery. When this phenomenon was first discovered, scientists thought the glow was due to thermal radiation, a result of the hot temperature of the water plume. But more recent measurements have shown that thermal radiation alone does not account for the entire amount of light from the deep-sea hydrothermal vents (a volcanic opening where super-hot water escapes from the depths of the Earth)—there has to be another source of light.

Scientists have suggested four potential mechanisms for this light source:

Crystallo-luminescence: As the hot, saltwater plume from the vent quickly cools in the surrounding seawater, the dissolved minerals crystallize and drop out of solution, giving off light.

Chemi-luminescence: Another mechanism is chemi-luminescence, in which chemical reactions taking place in the vent water release energy in the form of light.

Tribo-luminescence: A third mechanism is tribo-luminescence, in which light is given off as mineral crystals crack from the cold, or collide in the erupting plume.

Sono-luminescence: The fourth possible mechanism is sono-luminescence, in which light is given off as microscopic bubbles in the hot plume collapse.

What are **volcanogenic massive sulfide deposits**?

Volcanogenic massive sulfide (VMS) deposits are the remnants of hydrothermal vents (a volcanic opening where super-hot water escapes from the depths of the Earth) and sulfide mounds formed in the ancient oceans—some of which can now be found on land. They are large areas containing precipitated sulfide minerals (minerals that were separated from a solution), and are mined for elements such as copper, zinc, lead, silver, and gold. Equally important, certain VMS deposits contain fossils, which give scientists clues about the organisms that lived in and around these ancient vents. In particular, the Yaman Kasy VMS deposit, located in the Ural mountains of Russia, has yielded fossils from three different areas. In addition, fragments of black smoker chimneys were found there in 1995. This deposit dates from the Silurian period, approximately 430 million years ago.

Do **volcanoes and earthquakes** have anything in **common**?

Yes, volcanoes and earthquakes have one major thing in common: They both most often occur along plate boundaries. In addition, many times a volcanic eruption will be accompanied by an earthquake—before and/or after the eruption.

What is an **earthquake**?

An earthquake is the shaking of the earth, caused by the movement of the Earth's crust. The shaking is caused by the release of energy—in the form of seismic waves—as rock suddenly breaks or shifts under stress.

What are **seismic waves**?

Seismic waves are produced by the energy released by earthquakes; the waves radiate from the quake's epicenter (the point on the surface of the Earth directly above the focus of an earthquake). P-waves, or primary waves, move in a back and forth direction; they are the first waves produced by a quake and travel the fastest, reaching the other side of the world in about 20 minutes. S-waves, or secondary waves (also called shear waves), move side to side; they can pass through solids, but not liquids. Love and Rayleigh waves travel at the Earth's surface, like rolling ocean waves; these surface waves are responsible for most of the structural damage caused by earthquakes.

What is a **fault**?

Faults are fractures in the Earth's crust along which great masses of rock naturally move. They can occur between pieces of a crustal plate or as a boundary between two crustal plates. Not all faults are visible on the surface; many are found deep in the Earth's crust.

What do **earthquakes have to do with faults**?

When there is movement along a fault, it creates an earthquake. Some movement is so gradual and subtle that only sensitive scientific instruments can detect the activity. But if the rock movement along a fault is sudden and dramatic, it can cause a correspondingly powerful earthquake.

What is the **San Andreas Fault**?

The most famous example of a boundary between two slipping plates is the San Andreas Fault in California, in which a part of the Pacific plate slides by the North American plate. This 600-mile- (965-kilometer-) long and 20-mile- (32-kilometer-) deep fault is the "master fault" in an intricate network of faults that runs along the coastal region of California. The movement along the San Andreas measures less than an inch (some sources say up to 2 inches) a year—and scientists estimate that as a result of this movement, in 20 million years, the Pacific plate, on the western side of the fault will have slid northward so that Los Angeles and San Francisco will be neighboring cities.

What happens when an **earthquake** occurs **in the ocean**?

Usually when an earthquake occurs below the ocean floor, nothing significant happens—just the mild shaking of nearby lands. But certain marine earthquakes can cause a shifting of the ocean floor, especially along fault lines. When this happens, the displacement of earth can generate a seismic sea wave, called a tsunami—from the Japanese words *tsu* (meaning "harbor") and *nami* (meaning "waves"). These are gigantic and potentially damaging waves.

How are **earthquakes measured**?

Earthquakes, either in the oceans or on land, are measured using the Richter Scale, developed by American seismologist Charles Richter (1900–85) in 1935. An earthquake's intensity (magnitude) is measured on a scale of 1.0 (barely detectible) to 9.0 (extremely destructive). But the scale is not linear; instead, each unit represents an exponential increase: An earthquake that measures 7.0 is ten times more powerful than one at 6.0. Most earthquakes do not reach above 8.8. But unofficially, a 1960 earthquake in Chile reached a magnitude of 9.6—and spawned a deadly tsunami that hit Hawaii, Japan, and the Philippines.

What **major earthquakes** has the **United States** experienced?

There have been many major earthquakes in the United States, including Alaska. Most of these quakes occurred near, but not in, the ocean. And all but one occurred along the Pacific coast.

Richter Magnitude	Year	Location
8.3–8.6 (estimate)	1899	Yakutat Bay, Alaska (prior to statehood)
8.5	1964	Prince William Sound, Alaska
8.0–8.3	1811–12	New Madrid, Missouri
7.7–8.25 (estimate)	1906	San Francisco, California
7.1	1989	Loma Prieta, California
6.8	1994	Northridge, California
6.5	1971	San Fernando, California

What **major earthquakes** have occurred in **other parts of the world** in the **twentieth century**?

There have been many major earthquakes around the world in the twentieth century, most of them far from the oceans. These quakes had high magnitudes, and often caused extensive damage. But even lower magnitude earthquakes can cause great destruction and loss of life if it strikes a densely populated region where buildings are not designed or built to be earthquake-resistant. The extent of damage also depends on the number and the magnitude of aftershocks of a major quake.

Richter Magnitude	Year	Location
9.6 (not official)	1960	Chile
8.3	1994	Bolivia
8.2	1976	Tangshan, China
8.1	1985	Mexico City, Mexico
7.7	1990	northwest Iran
6.9	1988	Armenia
6.8	1995	Kobe, Japan

OTHER FEATURES OF THE OCEAN FLOOR

What are **seamounts**?

Seamounts are isolated volcanic mountains that rise from the ocean floor; the nearest equivalent on land is the singular volcano that rises above surrounding flatlands. (Even though seamounts are geologically separate from each other, they may occur in a chain.) Data gathered from more than 50 seamounts indicate that these features are remnants of now-extinct underwater volcanoes—and have a typical volcanic cone shape (though some can eventually fill in, becoming flat-topped). If the summit of a seamount has a small depression, this feature is called a crater; a larger depression is called a caldera.

Seamounts typically rise from 3,000 to 10,000 feet (914 to 3,048 meters) above the ocean floor, but are not high enough to poke through the water's surface. (There are exceptions to this, however, such as the Loihi Seamount, off Hawaii, which is expected to surface, creating a new island.) Seamounts have been found in all the oceans of the world, but the highest concentration is in the Pacific, where more than 2,000 have been identified and some of them are still active. The Gulf of Alaska has numerous seamounts rising from its floor.

What is a **guyot**?

A guyot is a flat-topped seamount; another name for this formation is a tablemount. American geophysicist Harry Hess (1906–69) discovered the guyot, naming it in honor of prominent geographer and geologist Arnold Guyot (1807–84). Seamounts are normally present near areas of volcanic activity, where there is a constant flow of molten rock (magma). If this magma accumulates to fill in the top of the seamount, its summit becomes level, producing a guyot.

What is the **Cobb Seamount**?

The Cobb Seamount is one of the most thoroughly explored seamounts: Due to the relatively shallow depth of its summit (only 124 feet, or 38 meters, below the water's surface), it receives ample light, which has

Scuba diver works near the Loihi Seamount, a new volcano forming in the waters off Hawaii. *CORBIS/ Roger Ressmeyer*

allowed scientists to examine it. It was discovered in 1950, and is one of a chain of seamounts that extends from the northern Pacific into the Gulf of Alaska. The Cobb Seamount is located approximately 270 miles (434 kilometers) off the coast of Washington, and rises nearly 9,000 feet (2,743 meters) from the ocean floor. Most of the seamount's summit, covering almost 23 acres, has been mapped by divers.

What is the **Axial Seamount**?

The Axial Seamount is located in the Pacific Ocean about 300 miles (483 kilometers) off the northern coast of Oregon. It is at the intersection of the Cobb Seamount chain and the Juan de Fuca Ridge—a seismically active mid-ocean ridge in the eastern Pacific that is spreading apart to form new oceanic crust. The Axial Seamount rises some 4,500 feet (1,372 meters) from the seafloor and its peak is still nearly 4,000 feet (1,219 meters) below the surface—but it's growing. The volcano is currently active, with more than 8,000 small eruptions detected during January 1998. In addition, glassy shards of rock have been brought up, indicating that magma is seeping or erupting through the crust.

Why are **seamounts important**?

Studies in the South Pacific Ocean south of Tasmania (an island off Australia) have found large numbers of fish and invertebrates, some new to science, living near seamounts. Among the more interesting finds were deep-sea coral reefs that are home to unique species, such as bamboo corals. And in one sampling, scientists found 259 species of invertebrates—such as coral, crabs, and sea stars—living on the seamounts, along with 37 species of fish. Approximately one third of the invertebrates were previously unknown, and up to 40 percent of these new species were thought to only occur on seamounts in this region. These findings indicate that seamounts are important, in global terms, as habitats for many unique and previously unknown species of flora and fauna.

What is an **ocean trench**?

An ocean trench is a very long, deep, narrow, V-shaped valley found near the edge of the continents, or close to island chains in the oceans. They form at subduction zones, as one continental plate slips under another plate, forming a deep depression on the surface of the ocean floor—the deepest points in the surface of the Earth. Ocean trenches usually occur parallel to the continents and island chains, and are areas of heavy seismic and volcanic activity. So far, explorers have identified 22 trenches—18 in the Pacific Ocean, 3 in the Atlantic Ocean, and 1 (the Java Trench) in the Indian Ocean. Among the more well-known trenches are the Tonga, Puerto Rico, and Mariana.

Major ocean trenches can exceed 18,000 feet (5,486 meters) in depth. In width, they vary from 10 to 22 miles (16 to 35 kilometers): The Tonga

Trench, located between New Zealand and Samoa, is the narrowest (and straightest); the Kurile Trench between Japan and Kamchatka is the widest. Their length can also vary widely: The Japan Trench is only 150 miles (241 kilometers) long and is the shortest, while the Peru-Chile Trench off the west coast of South America is nearly 1,100 miles (1,770 kilometers) long.

What is the **deepest trench in the ocean**?

The deepest trench in the ocean is the Challenger Deep, in the Mariana Trench, located in the western Pacific Ocean near the Mariana Islands (east of the Philippines). It was first measured in 1899; the latest measurement places its deepest point at 38,635 feet (11,776 meters), or approximately 7.3 miles (11.8 kilometers) below sea level. To compare, the tallest point on Earth, Mount Everest, measures about 29,022 feet, 7 inches (8,846 meters) above sea level.

What are the **major ocean trenches**?

The major trenches and their depths are:

Name	Ocean	Approximate Depth (feet / meters below sea level)
Mariana Trench	Pacific	38,635 / 11,776
Tonga Trench	Pacific	35,505 / 10,822
Japan Trench	Pacific	34,626 / 10,554
Kurile Trench	Pacific	34,587 / 10,542
Mindanao Trench	Pacific	34,439 / 10,497
Kermadec Trench	Pacific	32,963 / 10,047
Puerto Rico Trench	Atlantic	30,184 / 9,200
Bougainville Deep	Pacific	29,987 / 9,140
South Sandwich Trench	Atlantic	27,651 / 8,428
Aleutian Trench	Pacific	25,663 / 7,822

What is the **abyssal plain**?

The abyssal plain, or the abyssal floor, is a flat, relatively featureless area on the deep-ocean bottom; it is found next to the continental slopes. Abyssal plains lie at depths of about 6,560 feet (2,000 meters) to more than 19,680 feet (6,000 meters); the average depth is approximately 13,000 feet (4,000 meters). Here, the water temperature is near freezing, the pressure is immense, and there is neither sunlight nor changing seasons. Only a few living organisms have adapted to life in these inhospitable regions.

There are many abyssal plains across the oceans, including the Somali Abyssal Plain off the east coast of Africa, the Hatteras Abyssal Plain off North Carolina, and the Great Bight Abyssal Plain south of the Australian continent.

How were the **abyssal plains created**?

The abyssal plains are sometimes thought of as the true ocean floor; they have one of the smoothest surfaces on Earth, with less than 5 feet (1.5 meters) of vertical variation for every mile (1.6 kilometer). These smooth, level plains are a result of a steady and ongoing deposit of sediments, which fill in the nooks and crannies of the rough ocean floor. The sediments come from many sources; they can be washed down the continental slope by turbidity currents racing down submarine canyons, or they can drift down from the ocean waters above. And they can consist of materials from fine particles to the remains of marine life.

What are the **abyssal hills**?

Abyssal hills are one of two major features (plateaus are the other) on the ocean's abyssal plains. They are commonly low-relief features, usually found seaward of abyssal plains, or in basins isolated by ridges, rises, or trenches. The average height of the hills is 330 to 660 feet (100 to 200 meters) and the average diameter is about 33 feet (10 meters). Approximately 80 percent of the Pacific Ocean floor, and about 50 percent of the Atlantic Ocean floor, is covered by these hills.

What are **submarine plateaus**?

Submarine plateaus are one of two major features (abyssal hills are the other) on the ocean's abyssal plains. They are broad and more or less flat-topped features, usually more than 660 feet (200 meters) in height, though they vary in height and depth. Some plateaus are volcanic in nature; less frequently, they are the accumulation of deposited sediments. Submarine plateaus occur at various depths.

What is the **Kerguelen Plateau**?

The Kerguelen Plateau is a huge feature in the Indian Ocean; it measures about one-third the size of the United States and lies more than a half mile (less than a kilometer) under the surface of the water. This 100-million-year-old plateau was recently studied, as it is an example of a unique geological feature—a large igneous province (LIP). It may also be the result of one of the largest and longest-lived volcanic events on the Earth.

An LIP originates where magma (hot, liquid rock) wells up from deep beneath the surface of the ocean floor, and deposits molten rock. Some scientists believe large igneous provinces, which are the result of the largest volcanic events on Earth, may have affected the planet's past environment by altering ocean circulation, climate conditions, and sea levels. Apparently, such volcanic episodes were relatively common between 50 and 150 million years ago, but have been more infrequent during the past 50 million years. The depth of these plateaus have prevented their exploration: Until recently, they were almost inaccessible. But new technology has enabled their exploration.

AN OCEAN OF LIFE

How is **ocean life categorized**?

Marine life is classified into three general groups (this is only one of many ways to classify marine life):

benthos: These organisms live on the seafloor. They include corals and sponges (which are permanently fixed or immobile on the ocean bottom); snails and crabs (which scurry and creep along the ocean floor); and animals that burrow. This group also includes larger seaweeds, barnacles, and sea squirts.

nekton: These are swimming organisms, such as fish, that move freely and migrate from one place to another. They are able to swim in any direction, regardless of the ocean's currents. This group includes squids, herring, and many other familiar animals.

plankton: These are the smaller drifting and floating organisms—animals (zooplankton) and plants (phytoplankton). They usually do not have the ability to move from place to place with locomotion, but they may be transported by the action of the waves or currents.

How are the **ocean waters categorized**?

The ocean can be divided into three major zones, which are defined by the amount of sunlight they receive.

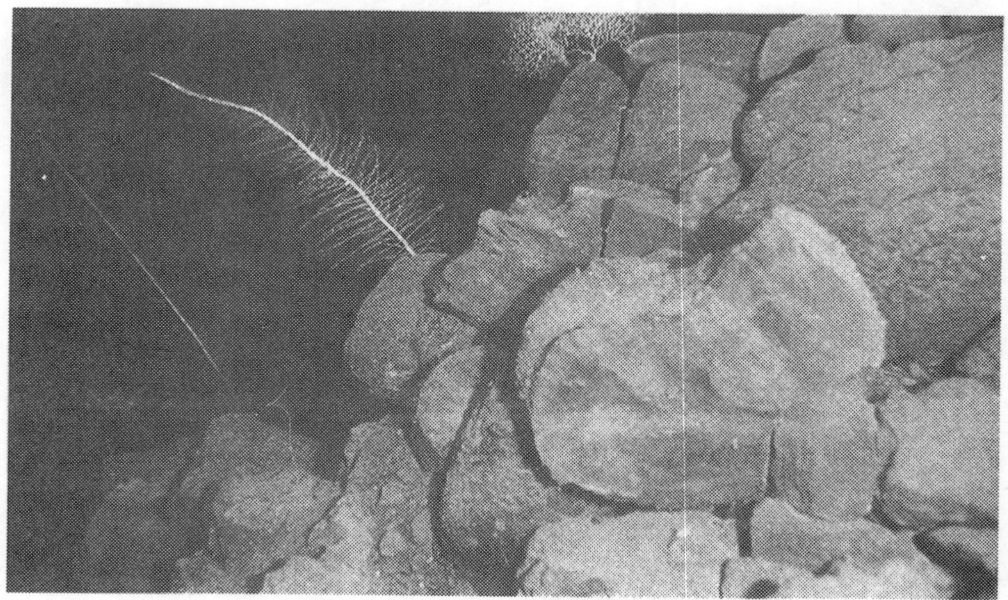

Sponges are categorized as benthos, which are those organisms that live on the seafloor. Here glass sponge grows atop pillow lava. *NOAA/OAR National Undersea Research Program*

The **euphotic (or epipelagic) zone** occurs at ocean depths to about 700 feet (roughly 200 meters), depending on the water's transparency.

Below that depth (700 feet, or 200 meters) to about 3,000 feet (roughly 900 meters) is the **disphotic (or mesopelagic) zone**, where very little light penetrates the water.

The area devoid of light, below about 3,000 feet (roughly 900 meters), which entails 90 percent of the space in the ocean, is called the **aphotic zone**. It is subdivided into the bathyal, abyssal, and hadalpelagic zones.

Where is **most life** found in the **oceans**?

Most of the plant and animal life in the oceans is found in the relatively thin uppermost layer called the euphotic (or epipelagic) zone. This layer extends from the surface down to no more than about 700 feet (213 meters).

Why is the ocean environment thought to be more stable than the terrestrial?

Marine organisms have the same four basic needs as do organisms on land—they require food, water, air, and a place to live. Terrestrial conditions are somewhat unstable, with relatively rapid changes in vegetation (caused, for example, by forest fires or extreme weather conditions), air quality (caused by pollution), and water (floods or droughts)—thus altering the environment surrounding terrestrial organisms. In the oceans, the environment is much more stable, continuously providing food, air, support for organisms' bodies, places to live—and especially water (with all its minerals and gases, too)—for all the organisms.

How are **organisms distributed** in the **near-shore waters** and **open oceans**?

Along the coastlines, marine life is found in areas with the best available shelter and food—especially along the nooks and crannies of rocky shores and coral reefs. In the open oceans, marine life is present down to several miles (or kilometers). But within the water column, the populations of organisms are unevenly distributed: Most organisms live in the euphotic (or epipelagic) zone (the top layer of the ocean), or they migrate to this layer in search of food.

How many **different types of plants and animals** live in the **oceans**?

There are estimated to be more than 250,000 different types of plants and animals living in the oceans—and there are without question many more to be discovered, especially in the deeper parts of the oceans. Some scientists estimate the number of species is closer to 400,000; and others even say there are between 1 million and 10 million benthic (deep-ocean floor) species yet to be described!

What do the studies of **marine ecology and biology** entail?

Marine ecology and biology entail the study of the plants and animals in the ocean—and are concerned with the relationship between the plants and animals as well as their relationships with their environment. This includes the way the organisms adapt to the various chemical and physical properties of the seawater. Researchers in these fields examine such variables as pollutants, the natural movements of the ocean, light availability, and even the stability of the seafloor.

MARINE PLANTS

What does **flora** mean?

Flora is another word for the plant life on Earth. It is also used to describe the entire plant life of a given region or habitat, or plant fossils in a geological stratum (layer of rock). The word comes from Flora, the goddess of flowers and springtime in Roman mythology. The plural of flora is florae or floras.

What is **photosynthesis**?

Photosynthesis is the process by which plants (and some single-celled flagellated organisms) convert inorganic carbon dioxide (found in the atmosphere), water, nitrite ions, and phosphate ions into useable sugars and amino acids by using the energy from sunlight. Plants use the process of photosynthesis almost everywhere on land. Photosynthesis also can occur in the ocean, in the euphotic zone, or in depths up to 700 feet (213 meters).

What is a **plant**?

Contrary to popular belief, the photosynthetic process does not define an organism as a plant; in fact, it is difficult to define a plant. These organisms vary widely in shape, size, and color, and range from independent, single-celled algae to the specialized cells that make up the multicellular plants we grow in our gardens. But there is one key microscopic feature common to every plant cell: An inflexible cellulose wall on the outside of the plant cell membrane.

What are living **cells**?

Cells are the tiny units that make up the majority of living organisms. About 60 to 65 percent of the cell is water because water is a perfect medium for biochemical reactions to take place. Cells are primarily composed of oxygen, hydrogen, carbon, and nitrogen. Within the cell, the most important organic materials are the proteins, nucleic acids, lipids, and carbohydrates (or polysaccharides). Certain cellular structures, all formed from a combination of these organic compounds, are also present in the cell.

What are the **two types of living cells** on Earth?

The two types of living cells on Earth are: the prokaryotes and eukaryotes. One of the greatest differences between these cells lies in their DNA: The prokaryotes' DNA are single molecules in direct contact with the cell cytoplasm (the cytoplasm is the living substance of the cell, excluding the nucleus); the eukaryotes' DNA have more than one molecule and the molecules can be diverse. Plus, the DNA of eukaryotes is found within a nucleus that is separated from the cell's cytoplasm by an envelope (a membrane); certain eukaryotes also are further divided by additional internal membranes. (The prokaryotes do not have such internal membranes.) In short, prokaryotes are simple cells and the eukaryotes are complex cells.

What are some examples of **prokaryotes** and **eukaryotes**?

The prokaryotes (simple cells) include bacteria and cyanobacteria (once referred to as blue-green algae). The eukaryotes (complex cells) include all plant (including algae) and animal cells.

How do **plant cells differ** from **animal cells**?

Plant cells have a rigid cell wall, a large vacuole (a fluid-filled pouch), and chloroplasts that are used for the synthesis of glucose (in which light energy from the Sun is converted to chemical energy—and finally into glucose). None of these features are found in animals cells.

Where and how did the **first plants evolve**?

Scientists believe that the first plant-like organisms evolved in the oceans about 3 billion years ago. To compare, the oldest known life is thought to be about 4 billion years old, in the form of small bacteria-like organisms. (However, to date, no such life has been discovered in the fossil record; the oldest known fossil evidence of life is about 3.75 billion years old.)

No one knows how organisms developed photosynthesis. Some scientists believe that certain early bacteria developed into chloroplasts—the special "organs" of a plant cell that carry out photosynthesis—which then released oxygen as a waste product, eventually producing the oxygen in our atmosphere.

What **sequence of events** led to **modern flowering plants**?

It is thought that the various plants evolved in the following way (and as usual with science, the discovery of more fossil evidence may change some of these dates in the future):

4 billion years ago: Single-celled life developed in the oceans.

3 billion years ago: The process of photosynthesis, with oxygen as a byproduct, probably evolved in some microorganisms in the ocean—eventually producing an oxygen-rich atmosphere. About 1.5 to 2 billion years later, the planet's protective ozone layer formed.

600 million years ago: There was an explosion of animal life—which evolved and became increasingly diverse, plants also flourished.

470 million years ago: Certain plants made the "first steps" onto land, adapting to conditions around the edges of tidal pools, called the land-water interface. Since the first land animals had not yet evolved, these plants were the first life on land.

430 million years ago: Plants with roots, stems, and leaves evolved; they are called vascular plants.

420 million years ago: The first true plants, including ferns, flowerless mosses, and horsetails, evolved on land and rapidly filled many ecological niches (environments). About 70 million years later, ferns developed seeds.

145 million years ago: The first flowering plants evolved on land; about 50 million years later, they dominated the land.

What **adaptations** did **marine plants** make in order to **move to land**?

The major change marine plants needed to make so they could move to land was to develop a new way of obtaining and keeping water, since they would no longer be submerged in the sea. Evaporation was also a problem, and plants had to develop a way to transfer necessary gases.

To solve these problems, various plants evolved many "new" mechanisms over millions of years. The one we are all familiar with is the seed—a waxy, waterproof cuticle that covers the embryo of the plant, preventing excessive loss of water. Some plants developed protective cells around the reproductive organs in order to minimize the loss of water. Still others developed spores, reproductive cells that develop directly into full-grown plants without having to be fertilized first.

What are some examples of **modern marine plants**?

There aren't as many modern marine plants as there are marine animals. But there are several that most of us recognize—including algae, marine grasses, and mangroves.

Do **marine plants** resemble **land plants**?

No, in the ocean, most plants do not look like the plants found on land. Most marine plants do not have stems, leaves, or roots; many cannot even be seen without the use of a high-powered microscope. But there are similarities: Marine plants make their own food through photosynthesis, just like land plants.

What are **algae**?

The great bulk of marine plant life consists of algae (the singular is alga). They are structurally simple single- to multi-celled organisms that carry out the oxygen-producing process of photosynthesis, and are found in both the oceans and in freshwater. You can see this type of organism by filling a clear jar with pond water and placing it on a sunny windowsill; or place a pail of rainwater outside. After several days, the water will look cloudy and green. This green water is filled with thousands of single-celled algae using the sunlight to carry out photosynthesis—similar to land plants that use photosynthesis to produce food in order to survive. Thus, algae are self-supporting and can live wherever there is the right balance of light, oxygen, carbon dioxide, and water.

There are numerous types of algae, named for their color, including blue-green, green, red, and brown. They are very diverse and are nonvascular (vascular plants, mostly on land, have vessels that conduct fluids up and down the plant; algae do not have these vessels). The free-floating algae are known as phytoplankton (plant plankton), and are usually single cells. They are distinguished from other marine plants by their reproductive processes.

How **big** are **algae**?

Contrary to popular belief, not all algae are small. Most algae are single-celled organisms, some as small as 1 to 2 micrometers in diameter (a micrometer is 1 one-millionth of a meter or 0.000001 meter—about 0.00004 inches); others are multicellular organisms, such as pond scum, seaweeds, and most green coatings on trees. On land, they can also live in symbiosis (mutually beneficial association) with certain types of fungi.

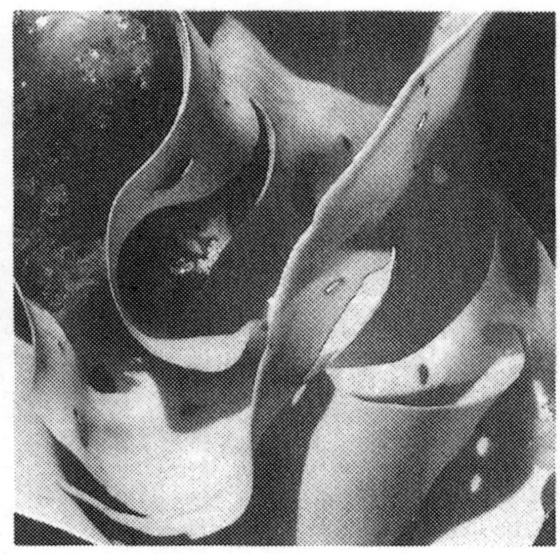
Ribboned seaweed photographed underwater off South Island, New Zealand. *CORBIS/Paul A. Souders*

The most complicated marine algae are the brown seaweeds. Bladderwracks, tangleweeds, and oarweeds are stuck fast to rocks in the ocean, and can grow yards (meters) in length. They usually have leaf-like fronds that float in the water, and contain chlorophyll that carry out photosynthesis.

Like land plants, the larger marine algae use photosynthesis, but the similarity ends there: Seaweeds do not have special roots that take in water; and they lack internal tissues like a xylem (land plants use these small internal tubes to carry water to their leaves). In addition, unlike most land plants, some seaweeds are able to adapt to long periods of drying.

Where are **algae** in the oceans?

The larger algae (macroscopic) are usually attached to a firm surface; they also grow on rocks in moving or stagnant water. In the oceans, they usually grow as seaweeds in the intertidal and subtidal zones as deep as 879 feet (268 meters), depending on the transparency of the water. The smaller algae (microscopic) are usually single-celled, free-floating organisms. They are a major part of the food chain in the ocean's upper layers.

What is **phycology**?

Phycology, also called algology, is the study of algae. Phycology is from the Greek *phykos* or "seaweed"; algology is from the Latin *alga,* or "sea wrack."

Why is the word **algae** often **misunderstood**?

The term is often misunderstood because there are so many types of algae. "Algae" merely refers to any aquatic organism capable of photosynthesis.

Why are algae important to the study of ancient life?

Fossilized single-celled algae have been found in rock that is more than a billion years old. It is thought that green and blue-green algae were some of the most successful plants that arose on the Earth. For such small single-celled plants to survive so well—and for more than a billion years so far, is amazing.

How do scientists **classify algae**?

Like all organism classifications, there is no one way in which scientists classify algae. Traditionally, the non-motile (mostly attached) forms of algae were thought of as plants, and the motile (mostly free-swimming) forms—even if they used photosynthesis—were considered to be both plants and animals. Now some scientists classify algae in multiple kingdoms; another classification puts most algae under the plant kingdom; and yet another classification puts algae in the kingdom Protista—except for cyanobacteria, which it places in the kingdom Prokaryotae.

The debate won't be over for a while. Research has suggested at least 16 lines in which groups of algae have a common ancestry, including the cyanobacteria, diatoms, and brown, green, and red algae. There will have to be more research into algae before any agreement will be reached as to classification.

What are the various **types of algae**?

There are numerous types of algae, including the following:

red algae (Rhodophyta): These algae are eukaryotic (made up of complex cells) and are mostly found in marine environments. They lack the chlorophyll-b that most other algae possess (they have chlorophyll-a instead), and there are special blue and red pigments within the algae. Because they have incomplete cell division when going through the reproductive process, they have "pit connections" between each cell; their actual reproductive history is very complex. The cell walls of certain red algae are also responsible for the production of two important polysaccharides—agar and carrageenan. Both have suspending, emulsifying, stabilizing, and gelling properties and are used in the production of some commercial foods.

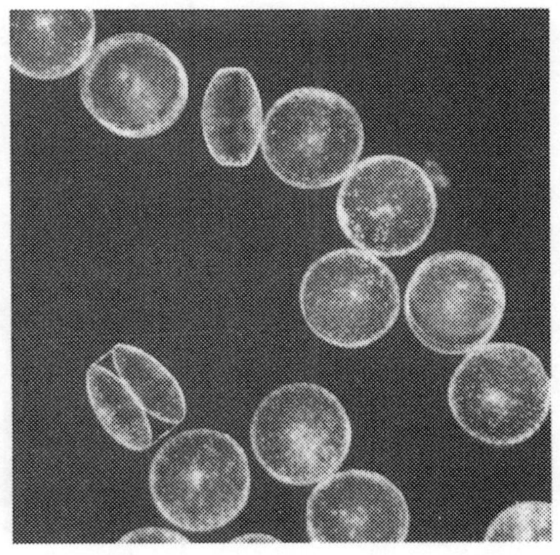

Diatoms are single-celled phytoplankton incapable of moving under their own power. They are extremely common and can be found in both salt- and freshwater. *CORBIS/Douglas P. Wilson; Frank Lane Picture Agency*

brown algae (Phaeophyta): These algae are found mostly in marine environments, but they lack the chlorophyll-b that most other algae possess. Instead, they have another type called chlorophyll-c, along with special photosynthetic yellow to deep-red pigments. Many of the brown algae grow to great sizes, and we know them as kelp. Kelp can reach up to 100 feet (30 meters) in length. Commercially, alginate is obtained from brown algae kelp, a polysaccharide used in the same way as agar and carrageenan (as suspending, emulsifying, stabilizing, and gelling agents in foods); other brown algae are used as sources of vitamins and fertilizers.

green algae: These algae are single-celled, or form cells in a group (colonial). They are similar to most plants, as they have both chlorophyll-a and -b, and also store food as starch. There are several green algae that live in the ocean. For example, one type has its cell walls infused with aragonite (a form of calcium carbonate); this algae makes an important contribution to a coral reef's formation and survival.

A small school of sturgeon feed on filamentous algae, which grew around a manmade reef of submerged pipes. *NOAA Sea Grant Program; Dr. James P. McVey*

dinoflagellates: These small organisms are an important part of the marine food chain; they are phytoplankton (small plant organisms) and have long flagella, or whip-like tails, that allow them to move up and down in the water column. They often have a multi-layered covering of cell material.

diatoms: These single-celled phytoplankton (small plant organisms) are found in both salt- and freshwater; they can even be found in moist soil. They are one of the most common types of phytoplankton. Diatoms do not have flagella (whip-like tails), and thus cannot move under their own power.

What are **cyanobacteria**?

Cyanobacteria are microscopic organisms; they are similar to bacteria because they lack a nuclear membrane (they are simple cells, or prokaryotes). They are also photosynthetic, and thus can manufacture their own food—but are classified by most scientists as bacteria, not as algae.

The cyanobacteria reproduce by binary fission (in which the cell splits into two identical cells), production and germination of spores, or by the breaking of multicellular filaments. Cyanobacteria are important to the growth and health of many plant species—especially those on land. This is because they are only one of a few organisms on Earth that can take inert atmospheric nitrogen and turn it into an organic form (as nitrate or ammonia). These organic forms are called "fixed" forms, which plants need in order to grow. In fact, fertilizers work in part by adding fixed nitrogen to the soil, where the plant roots absorb it. Certain types of cyanobacteria can also cause problems. For example, a species of *Lyngbya* can cause "swimmer's itch," a form of skin irritation that can develop from exposure to water infested with this cyanobacterium.

Why were **cyanobacteria once thought** to be **blue-green algae**?

Because they are photosynthetic and aquatic, cyanobacteria were once referred to as blue-green algae; but they are not related to any of the numerous eukaryotic algae. However, there really is such a thing as blue-green algae, and it is eukaryotic (has complex cells). Cyanobacteria are simple-celled relatives of bacteria.

Are all **cyanobacteria blue-green**?

No, not all cyanobacteria are blue-green in color. They can range from blue-green to purple. They get their name and part of their color from a bluish pigment called c-phycocyanin that is used to capture light for photosynthesis; they also carry a red pigment, c-phycoerythrin. Amazingly, cyanobacteria also carry chlorophyll-a, the identical photosynthetic pigment used by plants—but this green pigment is masked by the other two pigments.

There are also types of cyanobacteria that are red or pink—due to the pigment c-phycoerythrin—often found growing around sinks, drains, and even lurking on greenhouse glass. Another species is the *Spirulina,* a cyanobacterium that gives African flamingos their pink color.

Why were **cyanobacteria** important to the **evolution of life** on Earth?

Cyanobacteria were important to the evolution of life on Earth because of their relation to eukaryotes (complex cells): Sometime around the early Cambrian period (505 to 544 million years ago), cyanobacteria began to live within certain eukaryote cells, making food for their host in return for a place to call home. This is called endosymbiosis—and was the origin of mitochondrion structures (specialized cells containing energy-producing enzymes) that eventually evolved within eukaryote cells.

The other way in which cyanobacteria were important to the evolution of life on Earth dealt with the origin of plants: The chloroplasts—the place where the plants make food for themselves—are actually cyanobacteria living within plants' cells.

Seaweed can take many forms: Here it carpets the rocks along the shores of Cashel Bay near Connemara, County Galway, Ireland. ***CORBIS/Macduff Everton***

What are **stromatolites**?

Stromatolites are mostly ancient blue-green cyanobacteria; they were probably one of the earliest forms of life in the oceans. These organisms secreted lime, creating hard, large, layered, mushroom-shaped structures. Fossilized stromatolites have been found in many places; the oldest specimens were found near an area called North Pole, Australia. These ancient stromatolites, which released oxygen into the air, are believed to have played an important role in the build-up of oxygen in the Earth's early atmosphere. Stromatolites still live on Earth; one of the places they are found is in Western Australia, particularly in Shark Bay (in the Indian Ocean).

What are the most **common types of large algae** found along the **ocean shore**?

The most common types of large algae found in saltwater are mainly in the intertidal zones and clear waters less than 300 feet (100 meters) deep: they are green (*chlorophyceae*), blue-green (*cyanophyceae*), brown (*phaeophyceae*), and red (*rhodophyceae*) algae.

What is **seaweed**?

Seaweed is actually a larger form of algae. The most common are the brown and green algae that form brown- and green-looking seaweeds.

Is **seaweed** really **edible**?

Yes, one of the most popular seaweeds (actually red algae) is *nori,* which is an important part of many peoples' diet, especially in Japan. Various brown algae species include the popular *wakame, kombu,* and *hijiki*—all part of many people's diets, especially in Japan.

What is **sargassum**?

It is a seaweed that, unlike its algae relatives (which grow near the ocean shore), thrives in the open ocean. The sargassum weed—a brown algae—grows on the surface of the Sargasso Sea, a large tract of relatively calm water in the western Atlantic Ocean, near Bermuda. Here, sargasso forms a floating "island" of seaweed that covers an area measuring about two-thirds the size of the United States. This seaweed has air bladders located on its short stalk; they are also pelagic, or free-floating, and reproduce through a form of asexual reproduction (by breaking off pieces of the plants).

A remarkable community of animals lives in the floating mass of sargassum weed, including worms, bryozoans, shrimp, crabs, fish, and hydroids. Many creatures have adapted to the environment by imitating the color and sometimes texture of the seaweed in order to hide from predators. Other organisms use the seaweed for transportation: baby turtles, for example, often "ride" the algae out into the ocean.

Although 7 million tons of the seaweed lives in the Sargasso Sea, it is not thick enough to hinder navigation by humans. Occasionally, large chunks of the seaweed mass do break off and float away, washing up along the East Coast of the United States.

Are there any **other types** of **sargassum** seaweeds?

Yes, other sargassums are also found in waters along the East Coast of the United States; in all, there are about 15 different species stretching from

Kelp, a form of brown algae, can reach up to 100 feet (30 meters) in length and can form underwater forests or beds, such as the one the diver swims through here. ***NOAA/OAR National Undersea Research Program; W. Busch***

Maine to Florida. One sargassum, called *Sargassum filipendula,* is benthic (or attached to the bottom), and is found from Cape Cod southward.

What is **kelp**?

Kelp is a large group of brown algae (or brown seaweeds). They can grow into sizeable structures, sometimes reaching up to 200 feet (60 meters) in length; or they can form as branching mats. The largest are found off the western coast of North America where they grow on the rocky bottom, beyond where the waves break. Kelps along the Atlantic coast are smaller, reaching only about 10 feet (3 meters) in length and grow mainly below the tide line. They are often collected and used as a food source or for industrial raw materials. Bull kelp and feather-boa kelp are common types of the seaweed.

What are **kelp forests**?

Kelp forests are huge growths of the brown algae (or brown seaweeds). They can cover an ocean area of several miles with towering vegetation and thick canopies. These forests contain abundant marine life, including the popular sea otters, which dwell in the kelp beds along the central California coast.

Humans have also harvested the rapidly growing kelp beds, using large barges with paddle-wheel "mowers" to collect the seaweed (the blades of kelp rapidly grow back). Most of this kelp is processed into algin, a stabilizer and emulsifier used in many products, including paint, cosmetics, and ice cream.

What have scientists discovered after studying one of the world's **largest kelp forests**?

Just off the coast of San Diego is the Point Lomo kelp forest community. Researchers have studied the ecology of these kelp beds for years, trying to determine the various processes that affect the forest—including seasonal climate variations. They've determined that the effects of global climate episodes such as El Niño (a periodic warming of the waters off the west coast of South America) and La Niña (the following periodic cooling of the same waters) affect the kelp, while smaller, localized climate shifts do not. For example, during the 9-year study, *Macrocystis* species were not affected by the competition from other kelp species. But *Pterygophora californica,* an important understory (under the top canopy) species showed a reduction in growth and reproduction due to light-limited conditions and competition with *Macrocystis*; this occurred during La Niña periods when *Macrocystis* thrived. Conversely, when El Niño conditions led to poor *Macrocystis* growth, the understory kelps did much better.

PLANKTON

What are **plankton**?

Plankton are mostly tiny animal and plant organisms that float or weakly swim in the ocean's surface waters. Some plankton are single-celled while others are multi-celled—with some forming colonies. The word plankton comes from the Greek *planktos,* meaning drifting or wandering.

The organisms that make up plankton are very numerous and diverse; they can be grouped in many different ways. One division is based on the ability for photosynthesis; those plankton (mostly algae) that are capable of this process are called phytoplankton. Non-photosynthesizing (often termed animal) plankton are called zooplankton. The divisions of plankton are not made based on plant or animal characteristics—because some plankton can exhibit traits of both plants and animals.

Which **plankton** exhibit **characteristics of animals *and* plants**?

One example is the euglenoid, which is usually thought of as phytoplankton (plant plankton) and contains chlorophyll (for photosynthesis). But if the water changes and the light levels drop, these creatures will be forced to hunt for their food rather than manufacture it through photosynthesis. Thus, they bear characteristics of both animals and plants.

How are **plankton divided**?

One division (or classification) of plankton is by their size—from invisible to visible with the naked eye. The following table lists the names given to the various sizes of plankton and some common examples.

Name	Size (in microns*)	Example(s)
Ultraplankton	Less than 5	
Nanoplankton	5–50	protozoans (single-celled zooplankton)
Microplankton	50-500	invertebrate eggs and larvae
Mesoplankton	500 to 5,000	
Macroplankton	5,000–50,000	copepods
Megaloplankton	larger than 50,000	large jellyfish

*one micron is one micrometer, or one millionth of a meter (0.000001 meter)

Which **marine animals eat plankton**?

In healthy oceans, where there are few or no harsh pollutants or natural disasters, plankton are eaten almost as fast as they are produced. The common animals that feed on plankton include jellyfish, comb jellies, certain shrimp, herrings, anchovies, and even huge blue and gray whales.

Tiny anchovies, among the fish that eat plankton, are near the bottom of the food chain. *NOAA/OAR National Undersea Research Program*

Why are **plankton important** to ocean life?

Plankton are the foundation of the ocean food chain. All plankton are eaten by larger predators, such as fish or whales, or they die and sink to the ocean floor.

Who was **Viktor Hensen**?

German scientist Viktor Hensen (1835–1924) was the director of what is popularly known as the Plankton Expedition, a scientific expedition in 1889. The goal of this venture was to systematically categorize all the organisms in the sea. Hensen was responsible for naming the smallest organisms found by the expedition—the plankton.

What are **phytoplankton**?

Phytoplankton are the plant-type plankton living in the oceans and most other surface waters, such as freshwater lakes, rivers, and ponds. These small plants are found in water to about 100 to 130 feet (30 to 40 meters) deep. Similar to plants on land, the phytoplankton use photosynthesis—the process of using sunlight to make the food that gives them energy.

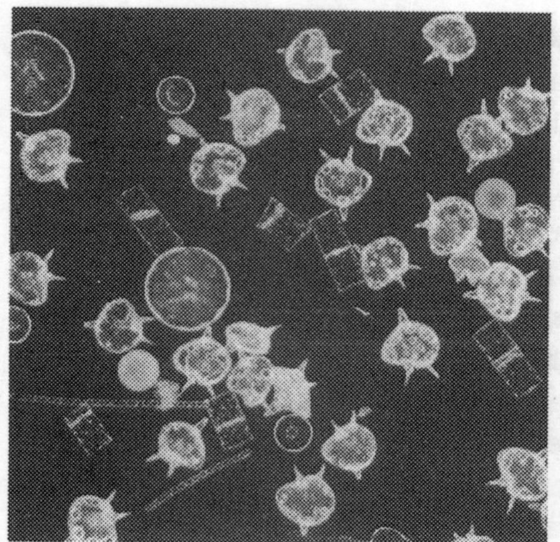

Dinoflagellates are phytoplankton with whip-like tails that allow them to move up and down in the water column. *CORBIS/Douglas P. Wilson; Frank Lane Picture Agency*

What **percent of marine plants** are **phytoplankton**?

Phytoplankton make up approximately 90 percent of the ocean's plants—and are the most prolific types of plankton (in other words, they far outnumber animal plankton, also called zooplankton). Individually, they are extremely small, but they are great in number!

What is a **dinoflagellate**?

A dinoflagellate is a phytoplankton that has a long flagella, or whip-like tail, that allows the organism to move up and down in the water column. Dinoflagellates often have a multi-layered covering of cell material; certain types are even armored with overlapping layers of cellulose plates. Some dinoflagellates are even bioluminescent, or light-producing; this group also includes those organisms that cause the red tides, when the sea water turns "red" from a profusion of dinoflagellates.

What is a **diatom**?

A diatom (Bacillariophyta) is a single-celled phytoplankton that is found in all types of water (ocean and freshwater) and even in moist soil. In the oceans, they are divided into two types: elongated or round (wheel-shaped) forms. They also have a type of pigment similar to brown algae, and their cell walls are made of silica.

Diatoms are considered one of the most common types of phytoplankton; along with dinoflagellates, diatoms are among the most abundant phytoplankton found in temperate coastal regions. Unlike dinoflagellates, diatoms do not have flagella (whip-like tails), and thus cannot move through the water under their own power. Instead, they have evolved certain ways to keep themselves afloat in the surface waters of the ocean, including specialized spines; certain species will also connect themselves to other diatoms to form long chains and increase their buoyancy.

Krill, a large zooplankton, are the foundation of the Antarctic marine food chain. *CORBIS/Peter Johnson*

What are **fossilized diatoms** called?

Accumulations of fossilized diatom shells are called "diatomaceous earth," a chalk-like deposit composed of silica. It is often used in filtration and as an abrasive.

What is a **common type** of **zooplankton**?

Copepods are the most abundant zooplankton in the surface ocean waters. They are the most numerous crustaceans in the world, with more than 6,000 species found in fresh and ocean water. They are about a fraction of an inch (just over 1 millimeter) in size and have varied body shapes.

Still other zooplankton are the larvae of much larger marine animals, such as corals and fish. Released in great numbers during the mating season, few of these tiny animals ever reach adulthood—as predators feed on the massive numbers of these tiny animals.

What are some **plant-eating zooplankton**?

Several types of zooplankton, such as copepods and krill, eat phytoplankton (plants). But this makes sense: The small size of these animals makes chewing on larger plants—or animals—impossible!

What is one of the **largest zooplankton**?

One of the largest zooplankton is the krill, a small, shrimp-like crustacean of the genus *Euphausia*; krill range in size from about 0.5 to 3 inches (1 to 7 centimeters). Many larger animals feed on these small zooplankton, including the filter-feeder blue whale, as do other whale species.

Why are **copepod plankton** so **important** to ocean life?

Copepod plankton are important links in many marine food chains. Not only do they feed on smaller microorganisms and phytoplankton, but they are also important food for larger animals.

What is a **carnivorous zooplankton**?

A carnivorous zooplankton called the arrow worm (or chaetognath) is known as the "tiger of the zooplankton." This organism feeds on animals, attacking and devouring prey as large or larger than itself. Most arrow worms range from about 0.4 to 1.2 inches (1 to 3 centimeters) long, and are shaped like darts. They are also free-swimming and have numerous chitinous teeth—and are themselves a prime target of larger species, which see them as a food source.

What happens to **plankton** that **die**?

While plankton are an important food source in the ocean, they are also prolific, which means there can be great quantities of them that are never consumed by predators. One place where plankton are plentiful is the Indian Ocean, where about 90 percent of the silt on the seafloor consists of the remains of plankton. Plankton blooms during the course of this ocean's summer monsoon, a time when the rains are plentiful, the weather is warm, and there are more nutrients for the organisms to feed on. Only the toughest parts (the resistant organic material or skeletons) of plankton are preserved when they die; this material is often broken up and slowly settles from the ocean surface to the ocean floor. One reason scientists are interested in this silt is because it is a record—a natural "climatological archive"—of the past several thousands of years, and therefore, worth studying.

Is there a connection between **plankton** and the **global climate**?

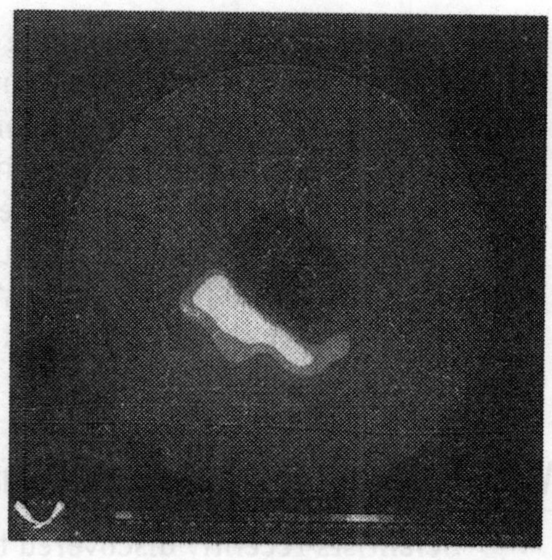

In October 1997 the National Oceanic and Atmospheric Administration (NOAA) released this graphic showing the "hole" in the ozone layer over the South Pole. (Also *see* the color graphic of this image, in the color insert.) Many scientists believe the enlargement of the ozone hole severely impacts the marine food chain. *AP Photo/NOAA*

Some scientists believe there is an important connection between plankton—particularly phytoplankton—and the global climate. These plant-plankton cycle (use) carbon and other elements in the atmosphere. If carbon dioxide increases in the atmosphere, the Earth's average temperatures could rise. Plankton remove carbon from the carbon dioxide in the air and use it in their respiration; they also store some of the carbon when they die and sink to the bottom of the oceans. But if certain nutrients, such as iron, nitrogen, or phosphorus are limited, the phytoplankton may not grow to their full potential—meaning less carbon would be absorbed from the atmosphere. Whether a reduction in the number of matured plankton would be enough to cause a major climate change is still unknown, but scientists are studying the possibility.

Does the **ozone hole** affect **plankton**?

Yes, some scientists believe that the ozone hole—the area in our atmosphere where the Earth's protective blanket has been destroyed—may have an important impact on plankton. This may especially be true in the Antarctic and Arctic oceans (around the South and North Poles, respectively). For several months each southern spring, two-thirds of the ozone shield protecting Antarctica and the surrounding ocean from the Sun's ultraviolet radiation is destroyed; up to half of the shield over the Arctic region is also lost for a short time each year. (These changes in the ozone layer are thought to be caused by chemicals from industrial and commercial emissions.)

Since plankton are susceptible to an increase in solar radiation, reductions in the ozone layer (or, put another way, enlargements in the hole in the ozone) affect these ocean organisms. When the ozone levels are low, scientists have noted a plankton loss of between 6 and 12 percent in the Antarctic Ocean. And since plankton are the bottom of the marine food chain, their reduction could affect all life in the oceans. Plankton can be thought of as the oceans' meadows: For example, krill feed off the plankton and marine mammals feed off the krill.

Why does **radiation from the Sun reduce** the number of **plankton**?

Scientists recently discovered that the reproductive cells of planktonic algae are several times as sensitive to the Sun's ultraviolet radiation as the organisms' mature cells. This means that the impact of solar radiation on plankton could be greatest in the spring—during so-called "blooms," when plankton reproduction is at its peak. If this proves to be true, marine plankton may be experiencing stunted growth. And because these animals are at the base of the food chain, an increase in the ozone hole could have serious effects on marine life.

Researchers found that the plankton's asexual spores (one way of reproduction among certain algae) are 6 times more sensitive to ultraviolet-B radiation from the Sun than are mature algae. This was determined by measuring the slowing rate of photosynthesis when the organisms are exposed to increasing radiation. Free-swimming gametes, which is the sexual means of reproduction of some types of plankton, were found to be even more susceptible: their photosynthesis fell by 65 percent after a 1-hour exposure to ultraviolet-B radiation—the equivalent of an ozone layer reduced by 30 percent—causing a reduction in plankton growth rate of about 17 percent.

MARINE ANIMALS

What does **fauna** mean?

Fauna is another term for the animal life on Earth. The word is also used to describe the entire animal life of a given region, habitat, or geological stratum (layer of rock). The word is derived from the Latin Fauna, the Roman goddess of nature. The plural of fauna is faunas or faunae.

How can **ocean animals** be described?

There is a wide range of animal life in the oceans, and many ocean-dwellers are unique and quite different from their land-dwelling cousins. Some are quite bizarre—having no legs, eyes, or ears. Other marine animals look and behave like plants, permanently attaching themselves to rock or the seafloor, where they siphon off oxygen and food from the water that passes by them.

Whether static or mobile, all marine animals obtain their food from their environment; they cannot produce their own food. They must gather and consume organic material in order to live. As on land, marine animals can be herbivorous (they eat plants); carnivorous (they eat meat, or other animals); or omnivorous (they eat both plants and animals).

What **animals** inhabit the **oceans**?

There are hundreds of animals inhabiting the oceans. The following lists only the major groups. (The following two chapters—Marine Mammals, Birds & Reptiles, which begins on page 261, and Fish & Other Life, which begins on page 303—describe most of these groups in detail.)

Sponges (phylum [order] Porifera)

Coelenterata
- hydroids (phylum Hydrozoa)
- jellyfish (phylum Scyphozoa)
- sea anemones (phylum Anthozoa)
- coral

Ctenophora
- comb jellies

Marine Worms

Bryozoans (phylum Bryozoa)

Mollusca (100,000 species; 7 classes; these are found close to shore)
- snails and other single-shelled (univalved) animals (Gastropoda)
- chitons (Polyplacophora)
- two-shelled (bivalved) mollusks (Bivalvia)
- squids and octopuses (Cephalopoda)

Arthropoda (75 percent of all animals—land and ocean)
- chelicerates (horseshoe crabs, spiders, and mites)
- insects
- crustaceans (crabs, shrimps, lobsters; nearly all marine arthropods are crustaceans)

Echinoderms
- sea stars
- brittle stars
- sea urchins
- sea cucumbers

Tunicates (phylum Chordata)
- sea squirts or ascidians

Fish (vertebrates; nearly 50 percent of the 40,000 known species of vertebrates are fish)
- Cartilaginous fish (phylum Chondrichthyes—sharks, rays, and skates; 10 percent of all fish)
- Bony fish (phylum Osteichthyes—tuna, cod, salmon, etc.; 90 percent of all fish)

Marine Reptiles (14 percent of the 40,000 known species of vertebrates are reptiles)
- crocodiles
- sea turtles
- sea snakes

Birds (sea and shore)
- puffins
- shorebirds (shallow water waders)
- egrets, herons, and ibis
- sea ducks
- gulls, terns, and skimmers
- cormorants
- pelicans
- gannets

What is the fastest marine animal?

The fastest known marine animal is the sailfish, *Istiophorus platypterus*. It has been clocked at speeds of 68 miles (109 kilometers) per hour.

frigatebirds
pelagic birds (such as storm petrels)
kingfishers
ospreys

Marine Mammals
whales (cetaceans)— baleen and toothed (toothed includes dolphins and porpoises)
seals, walruses, sea lions (Pinnipedia)
manatees or sea cows (Sirenia)
sea otters (Mustelidae)

What is **extinction**?

Extinction occurs when a species completely disappears from our planet. Extinctions of organisms have happened throughout the history of the Earth, often by natural occurrences, such as the impact of a space object (a comet, for example). Such an impact would cause material to be thrown high into the atmosphere, dimming the Sun and changing the local or global climate—depending on the size of the impacting body.

In modern times, the rate of extinction has dramatically increased—and most scientists believe there is a link between the increased extinctions and the growth of the human population. In other words, as the human population increases, more species become extinct—and this could happen at an increasing rate. By the year 2000, the world population is estimated to be approximately 6 billion. The rate of species extinction is currently at one species a day.

What is an **endangered organism**?

An endangered organism—plant or animal—is in immediate danger of becoming extinct. In most cases, the populations of these organisms are very low in number, and they need active protection (usually by a government) in order to survive. Unfortunately, even though animals are protected while on the endangered lists, their populations often decline. For example, the southern sea otter was driven to the brink of extinction due to hunting (for its fur) early in the twentieth century; it still numbers only in the few thousands even though it is protected. This is because these otters are very vulnerable to the effects of pollution, including oil spills. Still other endangered organisms make a comeback, such as the gray whale population, which was recently removed from the endangered species list.

Are there any **endangered marine animals**?

There are many endangered marine animals. The survival of these animals is at risk for many reasons, including the disappearance (or severe reduction) of their habitats, pollution, and even human activities (for example, marine mammals can get tangled in fishing nets or collide with boats). Marine mammals that are on the endangered list include southern sea otters; manatees; monk seals; and humpback, blue, fin, sei, right, and bowhead whales.

What is a **threatened organism**?

A threatened organism is one that is at risk and, if its population dwindles (or rather, continues to dwindle), it will be reclassified as endangered.

Are there any **threatened marine animals**?

Yes, there are several threatened marine animals. For example, the Stellar sea lion is now on the list of threatened marine animals because of its tremendous drop in population. Researchers and conservationists constantly monitor the populations of marine animals to determine if they are threatened; thus, the list changes often (some fall off of it while others are moved to the endangered list).

What are some of the **smallest animals** in the **oceans**?

Some of the smallest animals in the oceans are the marine protozoa—complete one-celled animals. These microscopic animals obtain food, breathe (respire), and eliminate waste products just like multi-celled organisms; they are found with plankton in the upper to deep-ocean layers. Protozoa can climb, crawl, swim, and scuttle, feeding on bacteria, other protozoans, diatoms (minute plants), and tiny organic debris. Three groups of protozoa are common in the ocean:

Sarcondinians: Sarcondina means "creeping flesh," which describes how these animals move. They have jellylike bodies, and almost ooze to form a "foot" to pull themselves along. The most common marine sarcondinians are the forams (found mostly in the deep oceans) and radiolarians (mostly in the upper layers of the ocean). Amoebas are also sarcondinians, but there are few marine species of amoebas.

Ciliates: These organisms—numbering more than 8,000 species—are covered with cilia, or hairlike structures. They use the cilia to move, eat, and breathe. Most of these organisms are solitary and free-swimming, but can be attached to colonies. They are often found between sand grains, where they eat bacteria and plant cells—and are food for other creatures that inhabit the same area.

Flagellates: Flagellates have whip-like tails called flagella. They move around using this projection; or they live in colonies attached to rock with a kind of flagella. Most of these animals feed on fine organic microparticles or bacteria.

THE FOOD CHAIN

What is a **food chain**?

A food chain is a complex, interlocking series of dependencies that life forms have on each other and on their environment. It is also the passing of nutrients from one organism, plant or animal, to another. No life form remains isolated and independent of its environment or of other life forms—all living things in the ocean, as on land, are bound together in an unending cycle of life and death. Numerous, small organisms form the base of the food chain; at the top are the largest organisms.

How does a **food chain work**?

A food chain normally begins with phytoplankton, which are the simple plant organisms that have the ability to manufacture food out of inorganic substances—they use the energy of the Sun's rays in the process of photosynthesis. These small organisms are eaten by herbivorous zooplankton (small, plant-eating animal organisms), who are themselves eaten by carnivorous zooplankton (small, animal-eating animal organisms). In turn, these zooplankton can be eaten by larger animals, such as fish, who are themselves eaten by larger fish. Organisms in the food chain that avoid being eaten eventually die, sink to the bottom, and sustain bacteria, which turn their complex, organic substances into simple inorganic nutrients; bacteria are also, in turn, eaten by larger organisms.

The nutrients produced by the bacteria eventually work their way back up toward the ocean surface, mostly through currents, to the phytoplankton; these nutrients are then used by the microscopic plants (phytoplankton) to begin the cycle again. In fact, the food potential of an area is often dependent on the rate at which these nutrients are recycled back into the top layers of the ocean.

Is there **only one food chain** in the ocean?

No, there are many individual food chains in the ocean. These chains can overlap and intersect to form complex food webs.

What are the **producers, consumers,** and **decomposers**?

In the oceans, the producers are the plants, as they make their own food through photosynthesis. The consumers are the animals, as they consume organic materials rather than making it themselves. And finally, the decomposers are bacteria, as they break down organic substances (dead organisms) into inorganic nutrients.

Sharks, at the top of the marine food chain, are predators and scavengers. ***NOAA/OAR National Undersea Research Program***

What happens to the **top predators**—the animals at the top of a food chain?

Animals that nothing else prey on are known as the top predators; eventually, even they die and sink to the bottom of the ocean. Their bodies are eaten by animals known as scavengers, such as crabs, lobsters, and sharks. Bacteria also attack the remains, decomposing them. This process breaks down their organic material into simple, inorganic nutrients, which are recycled to the ocean surface where they are eaten.

What can affect a **marine food chain**?

A marine food chain is directly affected by the populations (organisms) within its food chain and it can also be greatly affected by major external changes (since these have an effect on the organisms within the chain itself). Populations may be greatly affected by such activities as natural disasters (for example, earthquakes, floods, or volcanic explosions); weather and climate changes (for example, El Niño and La Niña ocean patterns); and the result of human activity (for example, chemical dumping, oil spills, or over-fishing).

Which **organisms** are the **basis of the ocean food chains**?

At the bottom of the food chain are billions of microscopic plants—mostly diatoms and other phytoplankton (tiny plant organisms)—the first links in most marine food chains.

Are there any **food chains** that **do not depend on sunlight**?

Yes, there are food chains that do not depend on sunlight—or on photosynthesis. The hydrothermal vents found in the ocean floor in volcanically active areas have organisms at the base of their food chain that depend on the warm, mineral-rich waters around the vents—not on sunlight—to make their own food.

What is a **food web**?

A food web is the complex intersections of food chains in a specific place and at a certain time. Scientists use the food chains to examine the relationships between plants and animals within the chain, between the plants, and between the animals themselves. The food webs are much harder to determine, as the relationships between species in various places (and in different food chains) is not always easy to interpret. Scientific interpretations of food webs are often educated guesses.

SMALL LIFE IN THE OCEAN

How are **ocean bacteria** described?

Bacteria (single-celled or noncellular tiny organisms) living in the oceans play a substantial role in what are called bio-geochemical processes—especially in the process of breaking down organic substances into inorganic nutrients. Bacteria are found from the coastal regions to the deep oceans.

There are many types of marine bacteria. For example, in the deep oceans, bioluminescent bacteria, or microbes that glow, are often found. And these bacteria may be indicators of pollution: Scientists have used these bacteria to determine the toxicity of water, because when these creatures are exposed to certain compounds, they darken.

Are **microorganisms** found in the **deepest part of the ocean**?

Yes, microorganisms (organisms, or bacteria, that can only be seen with a microscope) have been found in the deepest part of the ocean—on the floor of the Challenger Deep in the Mariana Trench (in the Pacific Ocean). In 1996, the submersible *Kaiko* dove to the Challenger Deep, collecting the first sample ever from such a depth. After growing organisms from the mud in a lab, researchers identified hundreds of species of bacteria, archaea, and fungi; most of the creatures were similar to those found living at other deep-sea sites. They also found bacteria that love the cold—a chilly 39° F (4° C)—surviving as resistant spores. The researchers also determined that these bacteria can handle pressures 1,000 times that at sea level. Scientists are working to identify the genes that help these microorganisms resist enormous pressures.

Is there a **connection** between **ocean bacteria** and **iron**?

Yes, scientists have recently discovered that oceanic bacteria are ruled by the iron content of the oceans. Some of these ocean bacteria need a diet of both iron and carbon to grow: Without iron, the microorganisms release most of the carbon into the atmosphere instead of using it to build new cells. In addition, scientists now understand that these bacteria compete with phytoplankton (plant organisms) for iron; the bacteria, gram for gram, contain twice as much iron as phytoplankton. But scientists have yet to determine the importance of any of these connections.

MARINE MAMMALS, BIRDS & REPTILES

DEFINING MARINE MAMMALS

What are **marine mammals**?

Marine mammals are air-breathing, endothermic (warm-blooded), vertebrate (having a bony or cartilaginous skeleton) animals. Their young develop within the mother; after babies are born, the mother cares for and feeds them.

How did **marine mammals evolve**?

About 350 million years ago (during the Paleozoic era), some amphibious creatures left the sea for the land. Some of these amphibians evolved into reptiles; some of the reptiles evolved into mammals. And some of these mammals eventually returned to the sea, evolving into marine mammals such as today's whales, seals, and manatees.

Marine mammals are relatively recent life forms, only appearing in the Cenozoic era, evolving approximately 65 million years ago. Seals and walruses appear to have developed approximately 54 to 65 million years ago, during the Paleocene epoch of the Tertiary period. Sea cows probably originated in the mid-Eocene epoch approximately 38 to 54 million years ago.

Whales are thought to have evolved during the Upper Eocene epoch, but this is highly debated. In fact, recently scientists digging in the Himalayan foothills found a fossilized jawbone from a new whale species named *Himalayecetus subathuensis.* This fossil indicates that whales may be older than previously thought—approximately 53.5 million years old.

What characteristics do **marine mammals** share with **land-dwelling mammals**?

Marine mammals have retained the main characteristics of land-dwelling mammals, such as a four-chambered heart, biconcave red blood cells, a diaphragm breathing muscle, and both the hard and soft palates of the mouth (oral) cavity. They also, like most of their land-dwelling cousins, bear their young alive, and feed them milk from mammary glands.

What **adaptations** did **marine mammals** develop to live in the oceans?

Over millions of years of living in the ocean, marine mammals adapted to their environment by developing special features that enable them to survive and flourish. These adaptations include powerful tails for propulsion in the water; lowered metabolic rate, which enables them to use less oxygen; a thick layer of fat, which insulates these organisms from the cold temperatures of the ocean; and milk that has a high fat and protein content, used to nourish their young.

What are the **main groups** of **marine mammals**?

The main groups (or orders) of marine mammals are the Cetacea, or whales, dolphins, and porpoises; the Pinnipedia, or seals, walruses, and sea lions; the Sirenia, or manatees (sea cows); and the Mustelidae, or sea otters, which are members of the weasel family (along with badgers, wolverines, and minks).

In early March 1998 this 40-ton, 60-foot (16-meter) blue whale was found floating in Rhode Island's Narragansett Bay. The rare mammal, one of the world's largest animals, was towed by the U.S. Coast Guard to a nearby beach where scientists (shown) could examine it. *AP Photo/Robert Button*

What is the **fastest marine mammal**?

The fastest known marine mammal is probably the sei whale, *Balaenoptera borealis,* reaching speeds of 35 miles (60 kilometers) per hour over short distances. Some scientists believe that the orca (*Ocrcinus orca,* or so-called "killer whale") may be a close contender; orcas may reach speeds of 42 miles (70 kilometers) per hour as they chase their prey.

What is the **largest marine animal and mammal**—and coincidentally the **largest animal** known to have lived on **Earth**?

The largest animal and mammal in the ocean—and thought to be the largest animal that ever lived on Earth—is the blue whale. This whale can reach a length of almost 100 feet (30 meters). It is estimated that approximately 6,000 to 10,000 blue whales remain in the world; they are found in the Atlantic, Pacific, and Indian oceans, and near the North and South Poles.

How **intelligent** are **marine mammals**?

According to behavioral biologist Edward O. Wilson, the smaller toothed whales (a group that includes several species, notably the orca) and dolphins (many of the approximately 80 species) are two of the most intelligent marine mammals. In fact, on a list of the ten most intelligent animals on Earth (excluding humans), the smaller-toothed whales come in seventh, while the dolphins rank eighth (two species of chimpanzees come in first).

How does the **breath-holding** capability of **humans** compare with that of **marine mammals**?

There is definitely a difference between the breath-holding capability of humans and certain marine mammals. While humans can and do swim, we are land-dwellers, and this is the environment to which we are best accustomed. Marine mammals, on the other hand, are accustomed to their watery environment, having made certain adaptations over time. The following list gives the average time certain mammals can hold their breath.

Mammal	Average Time (minutes)
human	1
pearl diver (human)	2.5
sea otter	5
porpoise	15
sea cow	16
seal	15 to 28
Greenland whale	60
sperm whale	90
bottlenose whale	120

How **deep** and **for how long** do some **marine mammals dive**?

The following table lists the maximum depth certain marine mammals can reach, and the duration they can stay underwater.

Mammal	Maximum Depth (feet / meters)	Maximum Time Underwater (minutes)
bottlenose whale	1,476 / 450	120
fin whale	1,148 / 350	20
porpoise	984 / 300	6
sperm whale	more than 6,562 / more than 2,000	75–90
Weddell seal	1,968 / 600	70

WHALES AND DOLPHINS

What are **cetaceans**?

Cetaceans are marine mammals of the order Cetacea, which includes all whales, dolphins, and porpoises. The order is divided into two suborders: toothed whales (Odontoceti) and baleen whales (Mysticeti). Dolphins and porpoises are toothed whales, as is the famous orca whale (or "killer whale"). The baleen whales (also called toothless or whalebone whales) include the humpback and gray whales.

What are the **characteristics** of the **cetaceans**?

The cetaceans have broad tails, tiny ears, and are hairless. They are streamlined to make swimming easier, and are propelled by the up and down movements of their horizontal tail flukes (the two lobes that protrude from their tails). They steer and stabilize themselves by using paddle-like flippers. These animals also have layers of blubber under their skin that are used to store energy, and for insulation from the cold.

What is the **origin** of the term **Mysticeti**, used to describe baleen whales?

The term Mysticeti is from the Greek word for "mustache." This is because the baleen on these whales somewhat resembles a hairy upper lip.

What is a baleen?

The baleen whales have a structure in their mouths called a baleen, a series of flexible keratin plates with hairy-like fringes that hang down from the whale's upper jaw. These long plates can be 2 to 12 feet (0.6 to 4 meters) long; the right whale is among the species having longer baleen. To feed, these animals merely open their mouths, letting water that is full of krill and other tiny plankton pour in. They then partially close their mouths, which forces the water out through the baleen. As the water passes through, the krill and other plankton are trapped on the inside of the baleen—which essentially acts as a filter. This food is scraped off the baleen by the whale's tongue, and is swallowed whole.

What are some **examples** of **baleen whales**?

The species that make up the baleen whales include the blue, finback, right, minke, gray, and humpback whales. The blue whale, *Balaenoptera musculus,* grows up to 98 feet (30 meters) in length; the finback whale, *Balaenoptera physalus,* can reach up to 82 feet (25 meters) long; the right whale, *Eubalaena gracialis,* reaches up to 56 feet (17 meters) in length; the minke, *Balaenoptera acutorostrata,* ranges between 26 and 33 feet (8 and10 meters) long; the gray, *Eschrichtius rubustus,* from 40 to 50 feet (12 to15 meters); and the humpback whale, *Megaptera novaeangliai,* grows up to 49 feet (15 meters).

What are some examples of **toothed whales**?

The toothed whales, or suborder Odontoceti (derived from the Greek for tooth, *odont*), includes more than 65 species, including dolphins and porpoises. Some of the better known examples of these whales are the sperm whale, *Physeter catodon,* 36–66 feet (11–20 meters) long; the pilot whale, *Globicephala melaena,* which can grow to 20 feet (6 meters) long; the har-

Bowhead whales (viewed from above), from the baleen suborder, are Arctic-dwellers. *NOAA*

bor porpoise, *Phocoena phocoena,* usually about 4–5 feet (1.5–2 meters) long; the bottlenose dolphin, *Tursiops truncatus,* ranging between 6 and 12 feet (2 and 4 meters); and the orca, *Orcinus orca* (also known as the "killer whale"), which can grow to 23 feet (7 meters) in length.

What do **toothed whales eat**?

The toothed whales are predators, grabbing their prey with their teeth, then swallowing them whole. The primary items on the toothed whales' menu are squid and fish, although some species will eat other marine animals.

How do **toothed whales locate** their **underwater prey**?

All toothed whales locate underwater prey by using their highly developed sense of hearing. Toothed whales have, over millions of years of evolution, lost their sense of smell but have retained fairly good vision.

But vision cannot always be used, especially when the water is murky, or when there is little to no light. To overcome these problems, toothed whales use their own form of sonar in a process known as echolocation.

Though they live in the water, whales are mammals and must breath air. Here, a whale is seen (white puff near center of picture) exhaling through its blowhole (its nose), situated on top of its head. ***NOAA/Commander John Bortniak, NOAA Corps (ret.)***

How do **whales breathe**?

Whales are mammals and must breathe air; unlike fish, they cannot extract oxygen from the water around them. In order to breathe, whales have a nose (called a blowhole) on top of their heads. This means the nose is exposed to the air as soon as the animal surfaces. The whale first exhales old air (the familiar "blowing"), then inhales. Its nose is then pinched closed, and the animal dives under the water, essentially holding its breath until it comes up again for air.

Baleen whales have two blowholes; toothed whales have one. The blowhole of the sperm whale is located on the left side of its forehead, so the spouting comes from the whale at a 45-degree angle, unlike the 90-degree top-spouting of other whales.

When a whale dives deeply, its bodily processes slow, cutting off the flow of oxygen to non-essential areas of the body. This conserves the amount of oxygen it has inhaled and increases the time needed between breaths.

Do **whales migrate**?

Yes, whales do migrate. For example, Korean gray whales journey from the Okhotsk Sea off the Siberian coast to the islands of South Korea. The California gray whale has the longest migration of any mammal, traveling along the West Coast of North America from the Bering Sea to Baja California in a yearly round trip of approximately 12,700 miles (20,434 kilometers). These whales journey south in the fall from their summer feeding grounds in the Bering Sea to shallow lagoons off Baja

A finback whale and her calf. At maturity, the finback, which is among the baleen (or toothless) whales, can reach up to 82 feet (25 meters) in length. *CORBIS/Judy Griesedieck*

California, where they give birth to their young during the winter season. In the spring, they make the return trip north to the Bering Sea.

Why do **whales migrate**?

Most whales seem to migrate for two main reasons: 1) the availability of food and 2) the mating and birthing cycles—both dependent on the season. For example, humpback whales migrate in the western Atlantic Ocean from around the Bahamas, north to Georges Bank, Cape Cod, and even Iceland during the summer to feed; they migrate back southward during the winter to breed. In the eastern Pacific Ocean, some humpbacks migrate up and down the North American coast and others migrate between the Hawaiian Islands (where they spend winters to breed) and the Gulf of Alaska (where they summer and feed).

Just how the whales navigate through the open ocean is unknown. Some scientists believe that the whales use the Earth's magnetic field as a guide, allowing the animals to orient themselves.

How do the **larger whales** compare in **weight and length**?

Here is how several of the larger whales, which come from both suborders (baleen and toothed) stack up in weight and length:

Whale	**Average Weight** (tons)	**Greatest Length** (feet / meters)
blue	84	98 / 30
finback	50	82 / 25
sperm	35	59 / 18
bowhead	50	59 / 18
right	50 (estimated)	56 / 17
humpback	33	49 / 15
sei	17	49 / 15
Bryde's	17	49 / 15
gray	20	39 / 12
minke	10	30 / 9

What is **unique** about the **gray whale**?

The uniqueness of the gray whale (a toothless whale) comes from its method of feeding. Like other baleen whales, the gray whale uses its baleen to trap food in its mouth as it expels seawater.

But this animal—an active, medium-sized whale of about 39 feet (12 meters), which is found in the North Pacific—goes one step further than just filling up its mouth with water and forcing it out. In captivity this whale has been observed swimming just off the bottom, then turning on its side and sweeping its head back and forth. This behavior disturbs any crustaceans resting on the bottom, causing them to rise. The gray whale then creates a suction by pushing its tongue against the bottom of its mouth while expanding its throat grooves. This pulls the prey into its mouth under the short baleen. Think of this as the gray-whale version of the vacuum cleaner, removing crustaceans from the ocean bottom!

Although this behavior has only been observed in captive whales, there are several factors that suggest this is also a typical mode of feeding for these animals in the wild. The gray whale's baleen is usually shorter on one side of its mouth—and that side is often relatively free of barnacles,

The gray whale is a North Pacific–dweller: Here one of three that got caught in ice in the Beaufort Sea and became the subject of an American-Russian rescue effort. ***NOAA/Office of NOAA Corps Operations***

suggesting that this feeding behavior is not only normal for these animals, but that they adapted physically to make this suction trick possible.

What has **happened** to the **right whale**?

The right whale (a toothless, or baleen, whale) is the most endangered of the big whales—and may be at the brink of extinction. Worldwide, their numbers have dwindled to between 350 to 400. Many right whales, indigenous to the North Atlantic Ocean, migrate as far south as Florida during the Northern Hemisphere's winter, and back toward Canada in the summer. But the whale counts during these migrations indicated that only a few whales have borne calves in recent years.

There have been attempts to protect the critical habitats of the right whales. For example, in 1994 the New England Aquarium began operating an early warning system in the southeastern United States to let ships know the relative positions of calving right whales. Every winter, the aquarium researchers conduct daily flyovers of the right whale's calving grounds; they then relay the information to the Coast Guard,

Beluga whales are among those species that swim in pods. (Here at least ten are seen from above.) *NOAA/Captain Budd Christman, NOAA Corps*

Navy, and local harbor pilots—who in turn, relay the information to military and commercial vessels.

Do **whales** swim in **groups**?

Yes, most whales travel in groups called pods, with each species varying in its mode of travel within the pods. For example, some whales, such as the orcas (so-called "killer whales"), remain in tight family units; others swim in large pods, such as the bottlenose dolphin, which lives in groups of up to 12; while still others, such as gray whales, are often seen swimming alone or a mother is seen swimming with her calf or calves.

Do **humpback whales sing**?

Yes, humpback whales (which are toothless) do produce "whale songs," but scientists don't know how the sounds are generated. Humpbacks have no vocal cords, yet their sounds (classified as true songs) cover the widest frequency range of all the whales. And as strange as it may seem, when the whale sings, no air is released into the water—unlike humans who have to force air through their vocal cords in order to sing.

Most whale "singers" are males, with most of the songs sung at the breeding grounds of the animals. Many researchers believe the songs are linked to the courtship of the animals—either to attract a mate, advertise the mammal's availability, or establish a territory. The songs travel fast and great distances, too, since water conducts sound 3.5 times faster and 4.5 times farther than in air: Some humpback songs can be heard miles away—even as far as 100 miles (161 kilometers) if the frequency of the song is low enough.

Are the **humpback whales in danger**?

Yes, humpback whales are in danger. They are listed as an endangered species, and are thus protected worldwide. Less than a century ago, there were about 120,000 of this species of baleen whales; because of commercial whaling, their numbers eventually dwindled to about 10,000. Today, the numbers are not even close to previous numbers, and only a few scattered breeding sites remain. Not only is there the threat of whaling (not all countries have stopped), but there are also problems with pollution and human interference (mainly in the form of the humpbacks having mishaps with marine vessels).

What are **sperm whales** like?

Almost everyone's vision of a sperm whale comes from the famous tale of the sea told by American writer Herman Melville (1819–91) in his classic novel *Moby Dick*. (In this story, Moby Dick is described as a "great white whale," which is actually a white sperm whale, and he was the nemesis of Captain Ahab.) Sperm whales are toothed whales that have large, blunt noses, and squared-off heads. In fact, the head of this whale is about one-third the animal's entire length. They grow to weigh as much as 35 tons and can be as long as 59 feet (18 meters). They also have the largest brains of any mammal on Earth (maybe this is why Melville chose this particular whale for his story). They live in all the oceans and are the most numerous of the great whales.

A dolphin escort for the National Oceanic and Atmospheric Administration ship *Peirce* (1985). *NOAA; Personnel of NOAA Ship* Peirce

Why have **sperm whales** been **hunted**?

Sperm whales have been hunted extensively—in the past, and to a modest extent today—for a variety of reasons. They are a source of spermaceti, a white, waxy substance obtained from the oil in the animal's head; and ambergris, a material from the animals' digestive tracts. Both these substances are used in the manufacture of cosmetics and the spermaceti is also used for candle wax or long-lasting lamp oil. Other parts of the animal, such as the blubber, have also been valued—thus the sperm whale's decrease in numbers.

What is the **difference** between **dolphins** and **porpoises**?

Though perhaps indistinguishable in pictures, if you saw a dolphin and a porpoise side by side, you would notice the differences between these two toothed whales. These marine mammals—with about 40 species between them—differ in the structures of their snout and teeth. True dolphins have a beak-like snout and cone-shaped teeth; true porpoises have a rounded snout and flat or spade-like teeth.

How are **dolphins similar to humans**?

They may not seem like us, but researchers recently discovered that the dolphin genome (set of chromosomes) and the human genome are basically homologous—the same. There are just a few chromosomal rearrangements that changed the way the genetic material was put together. Because of this, scientists are now studying diseased dolphins and sporadic die-offs closely, trying to determine how these animals react to disease and to unfortunate exposures to pollutants. Some researchers wonder if dolphins will turn out to be the proverbial "canary

Killer whales (Orcinus orcas) come up for air in the icy waters of the Antarctic. *NOAA*

in the coal mine," signaling to humans when there are too many toxins in our environment.

What are "**killer whales**"?

Killer whale is another name for the orcas. These carnivorous mammals are the largest of the dolphins; the males may be as long as 30 feet (9 meters); females are smaller. Most orcas are black and white, although some have been reported to be all black or white; their young are black and orange. This toothed whale hunts in packs (pods) and is found worldwide.

Like many larger creatures of the sea, orcas have no natural enemies. But they do hunt other creatures—and only for food, not sport (the name "killer whale" is highly exaggerated, and they rarely attack humans). While they are known to eat birds, fish, and squid, they are also the only cetaceans to feed on mammals, including smaller dolphins, seals, and porpoises. And while in a pod, they may even attack a baleen whale.

What is **spyhopping**?

Orcas, or killer whales, exhibit a behavior called spyhopping. This is when each orca "stands" vertically—up to halfway out of the water. They do this to look around and see above the water.

What is **echolocation**?

Echolocation is the process used by certain animals (including bats) to determine the shape, size, and distance of an object; it is based on the sound waves generated and received by the animal. Scientists don't know how echolocation works in most marine mammals, such as the sperm whale. The most complete studies to date have been done on the smaller toothed whales, such as the dolphin. (It's much easier to study an animal that is not large enough to swallow you whole!)

Dolphins can generate a wide range of sounds underwater. Some sounds, such as chirps, moans, and squeals, are thought to be used in communication with other dolphins. Dolphins can also emit series of very short clicks, which are used in echolocation. These clicks can be generated at a rate of up to 800 times per second, and are thought to be emitted from the animal's blowhole—as an outgoing beam of sound pulses from an organ in the dolphin's forehead. (This organ is the noticeable "melon" on the head of the dolphin—a large, lens-shaped organ made of a waxy tissue and located between the top of the head and the upper jaw.)

When these out-going sound waves strike a target, such as a fish, some of the waves reflect back in an echo toward the dolphin. These reflected sound waves are received by the bony lower jaw of the animal, transmitted to the bone-enclosed inner ear, and converted to nerve impulses that are sent to the brain. There, the amount of time that elapsed between the out-going clicks and the returning echoes are used to determine the distance to the target.

The dolphin varies the rate of click generation, so that incoming echoes can be received in between out-going sounds. The animal uses low-frequency clicks to scan and higher frequency clicks to make more precise determinations. Using this sophisticated process, a dolphin can continuously determine the shape, size, and distance to an object underwater—as well as the direction of that object's movement.

Do any other **marine mammals** use **echolocation**?

Yes, in addition to the toothed whales, some baleen whales—such as the blue and gray minkes—also use echolocation. And this process isn't just limited to the cetaceans: Some pinnipeds, such as the California sea lion, Weddell seal, and walrus, also use echolocation.

How have **recent fossil finds** provided clues about **whale evolution**?

Scientists recently discovered a previously unknown genus and species of whales—a whale that has so far, not been named. The almost complete fossil was found by amateur fossil-hunters in a quarry on Washington's Olympic Peninsula, between Port Angeles and Clallam Bay. The bones were widely scattered and many of them were broken; they were found in an area that was a seafloor 28 million years ago, during the Oligocene period. This was also a period when whales were going through major evolutionary changes.

Scientists painstakingly reconstructed the animal—and discovered the uniqueness of the specimen. Based on this rebuilt skeleton, the whale was 15 to 18 feet (4.6 to 5.5 meters) long—about the size of a modern pygmy right whale, which weighs 3.25 to 4.6 tons and is among the smallest of the baleen whales. It was toothless, having baleen (plates along the upper jaw) to trap food while expelling ocean water from its mouth. Its blowhole was on its snout instead of the top of the head. In addition, the length of the arm bones and the way the rib bones attached to the backbone vertebrae more closely resembled land mammals. Since whales from more recent periods evolved bones that don't as closely resemble land animals, scientists are looking into the possibility that this whale represents a step in the evolution of the species.

SEALS, SEA LIONS, AND WALRUSES

What are **pinnipeds**?

Pinnipeds are a group of carnivorous, marine mammals. The word Pinnipedia, which means fin-footed (or those with "winged feet"), is

On land, sea lions move around on all fours. They make themselves at home along the Pacific coastlines of North and South America as well as New Zealand. *NOAA/Captain Budd Christman, NOAA Corps*

used to describe this group, since all four limbs of the pinniped were, through evolution, modified into flippers. Pinnipeds are carnivorous (meat-eaters). This group includes seals, sea lions, and walruses, all of which are very awkward on land, but are swift and agile in the water.

How can **seals, sea lions, and walruses** be **identified**?

Members of this group (Pinnipedia) can be confusingly similar in their appearance, but there are important differences that distinguish each. True seals (phocids) have no external ears and use only their forelimbs for motion on land. Sea lions, on the other hand, have ears and when on land these animals move around on all fours. Fur seals are an intermediate group between the true seals and the sea lions—they have ears, move on all four limbs, and have thick underfur. Finally, the walrus is really a close relative of the seal; it has no external ears, but it moves on all fours. These large animals are also known for their ivory tusks.

Walruses sunbathing on smooth rocks. Other than polar bears—and humans, who have hunted the animal for its ivory tusks and blubber—the walrus has no other predators. ***NOAA/Captain Budd Christman, NOAA Corps***

What are **walruses**?

Walruses are giant marine mammals related to the seals, but constitute a distinct family of their own. They are large, tusked pinnipeds that inhabit the cold, northern waters of the Atlantic and Pacific oceans, near the edge of the Arctic ice pack. Males are larger than the females, with an average length of more than 10 feet (3 meters); these marine mammals frequently weigh more than 2,500 pounds (1,135 kilograms). A layer of thick blubber insulates the walrus from its harsh, cold climate. The most prominent feature on walruses, both male and female, are their two prominent tusks—really enlarged canine teeth—which are used for territorial defense and courtship displays.

Although walruses are much larger, they do share many of the characteristics and body shapes of the seals. Like true seals (phocids), they have no external ears. These animals have whiskers and brushlike mustaches on their faces, and short reddish hairs on their hides. They feed on shellfish; when not searching the bottom for food, they spend most of their time hauled out on ice floes, resting and sleeping. Walruses are social animals, living in big groups dominated by the largest males, who collect harems of females. The only predators of this large pinniped are polar bears, who mainly prey on the young animals; humans also constitute a danger to walruses, which, for their ivory tusks and blubber oil, have been the object of hunters.

What are **sea lions** and where do they **live**?

Sea lions can be thought of as eared seals—they are related to the fur seal, but do not have the coat that has made their cousins the object of hunts. Sea lions are Pacific Ocean–dwellers. Among the varieties are the Stellar sea lion, which lives in Alaska and the Pribilof Islands (in the southeast Bering Sea); the California, which inhabits western North America and the Galapagos Islands; the South American, which can be found along the western coast of South America; and Hooker's sea lion, which lives in New Zealand.

What are **seals**?

The term seal is often used very loosely, to refer to any marine mammal in the order Pinnipedia, including walruses and sea lions. But scientists narrowly define a seal as only the phocid (from the family Phocidae), which is the "earless" seal. These seals do have ears, but they are internal; in other words, they lack the visible external ear of other animals. Further the "true seal" uses only its forelimbs for motion on land.

Of the roughly 17 species of seals, only 1 lives in warm water; the others live in cold and temperate waters. True seals are awkward on land, but agile in the water. Carnivores, they like to eat squid, fish, and assorted invertebrates.

How do **seals hunt**?

Recently, scientists attached a tiny video camera to a Weddell seal. The seal took a deep breath and dove 330 feet (100 meters) below the Antarctic ice sheet, stalking its dinner for about 20 minutes before resurfacing for air. The video of the Weddell seal stalking its prey revealed several interesting things: The seal has sharp eyesight—and hunts by sight

A Weddell seal pup; though the Weddell uses echolocation (sonar), scientists recently found that this animal has keen eyesight, which it uses to hunt for food. *NOAA/Commander John Bortniak, NOAA Corps (ret.)*

rather than sound as was once thought; the seal blows a blast of air from its nostrils to "encourage" prey hiding in an ice crevice to come out into the open; it can come within inches of its prey without being detected; and finally, even after traveling more than 1 or 2 miles underwater, seals can find their way back to a small air hole in the ice's surface.

What are some **true seals** and where do they live?

The following table lists some of the true seals—or phocids—and their habitats.

Common Name	Habitat
ribbon	Siberian coast
ringed	Arctic
larga	North Pacific
northern elephant	western North America
southern elephant	South America
Weddell	Antarctic

Common Name	Habitat
Ross	Antarctic
leopard	Antarctic
crabeater	Antarctic
harbor	Gulf of Maine

What are **fur seals**?

Fur seals, often called sea bears, belong to the same family as sea lions, both of which have external ears. They have a double coat of fur, with the underfur (or undercoat) being very dense and soft, and unfortunately, prized by hunters. Varieties include the northern, Alaska, Galapagos, Australian, and Antarctic fur seals.

What is the **largest pinniped**?

The largest pinniped is the elephant seal. The males can be more than 20 feet (6 meters) long; whereas the female measures about half that size.

What is an **elephant seal**?

An elephant seal is the largest pinniped—growing to 16 to 20 feet (5 to 6 meters) in length; some males weigh up to 7,500 pounds (3,405 kilograms). They come ashore only twice a year—once to molt (they shed their skin in a month) and once to mate. At other times, they are offshore, searching for food. They live an average of 20 years.

How are the **male elephant seals** distinguishable?

One of the main features of a male elephant seal is its large nose, or proboscis. This large structure on the front of the animal's face is actually a secondary sex characteristic: As the male grows and matures, the nose grows, too—becoming fully developed at about 8 to 10 years of age. The males use the nose for display purposes during mating seasons—the males holding their noses high in the air while bellowing challenges to each other.

The elephant seal is the largest pinniped; males can grow to more than 20 feet (6 meters) in length. *NOAA/Commander Richard Behn, NOAA Corps*

Are **elephant seals** near **extinction**?

Elephant seals are not quite near extinction, but they are struggling. During the 1800s when they were valued for their oil (as a light source for humans), elephant seals were hunted to near extinction. They recovered during the first part of twentieth century, but now they are threatened once again. The demand for the animals' oil is not the problem this time, but rather a danger is posed by the fishermen that ply the same waters where the elephant seals live and eat. The problem stems from the size of these animals: Males are up to 7,500 pounds (3,405 kilograms), and even the pups can weigh 300 pounds (136 kilograms) by the time they are weaned and on their own. To maintain this huge size, the animals must eat great quantities of fish and squid—the same food that the commercial fishermen collect. In other words, the elephant seal now vies with the fisherman for food.

To learn more about this marine mammal, scientists are conducting a decades-long study of the elephant seal population at the Peninsula Valdes (part of Argentina's Patagonian reserve), the fourth-largest elephant seal population in the world—and the only growing population.

SEA COWS, DUGONGS, AND MANATEES

What are **sirenians**?

The marine mammals that belong to the order Sirenia are the sea cow, dugong, and manatee. These aquatic creatures are herbivorous (plant-eaters).

What are **sea cows**?

Sea cows are marine mammals that grow up to 10 feet (3 meters) in length, have little hair, and rely on a layer of blubber for insulation. Their scientific name, Sirenia, is derived from the comely Sirens who, in the Greek poet Homer's *Odyssey,* lured Odysseus' crew into the rocks—a reference to ancient sailors who thought the sea cows were actually mermaids.

What is a **dugong**?

A dugong is a type of sea cow found off Africa and in the South Pacific—but mostly near Australia, where it is endangered. The dugong's closest terrestrial (land) relative is the elephant; it is also the only herbivorous mammal living totally in the ocean. Dugongs eat mostly sea grass, and use their short-bristled muzzle to dig and root in the shallows of bays and estuaries. They have two flipper-like forelimbs, but no hind limbs. The tail is broad—allowing the animals to propel themselves through water at speeds reaching about 13 miles (21 kilometers) per hour. Although they may seem slow, they are active—and some scientists believe the animal's intelligence is comparable to a deer.

What is a **manatee**?

A manatee—a thick-skinned and wrinkly marine mammal—is a close relative to the dugong, and is also a sea cow. In fact, the only major difference between the two is their habitat: The manatee is usually found along Florida shores, although their range once spread from North Carolina to Florida. Manatees perform a service to marine craft: They use

The lure of the sirens? The wrinkly and thick-skinned manatee belongs to the order Sirenia. *CORBIS/Brandon D. Cole*

their bristly muzzles to dig and eat the bothersome water hyacinth that can clog slow-flowing river channels. Like the dugong, the manatee is endangered.

Why are **manatees** and **dugongs** endangered?

One of the main reasons has to do with their appetites: Because they root for the sea grasses in estuaries and shallow bays, they encounter more recreational craft than do most ocean animals. A manatee comes up for air every 10 to 15 minutes, but the rest of the time, it is often lying placidly in the shallow water, just under the surface—where they are often hit by boats. In addition, they are very slow, and are not able to get out of the way of such craft. Thus, many of the animals are injured by propellers or fishing nets; they are also extremely susceptible to silt filling their shallow water habitats—not to mention pollutants and petroleum spills.

There are major efforts underway to protect the manatees and dugongs—especially by studying their populations and by not disturbing the animals' habitats. For example, manatees off the coast of Florida are closely watched. Many animals, usually captured as abandoned young or hurt animals, are tagged, released into the wild, and monitored by satellite—in hopes of keeping an eye on the number of calves born, so that population shrinkage and growth can be gauged. And in western Australia's shallow waters, there are laws that limit how close vessels and people are allowed to come to the dugongs: no closer than 330 feet (100 meters) for vessels and 100 feet (30 meters) for people.

SEA OTTERS

What are **sea otters**?

Sea otters are the smallest marine mammals; they are also the least changed in form from their terrestrial (land) ancestors. Sea otters are relatively large animals—the males often grow to 4.5 feet (1.35 meters) in length, including the tail, and weigh about 45 to 100 pounds (20 to 45 kilograms). They are closely related to the smaller river otters found in freshwater streams, and are members of the weasel family (the order Mustelidae).

Are there **different** types of **sea otters**?

Yes, there are three subspecies of the sea otter: the southern sea otter (found off the California shore), Alaskan sea otter, and Asian sea otter. The differences between these animals includes the shape of the animals' skulls and the size of the otters (the Alaskan sea otters usually grow the largest).

What do **sea otters eat**?

Sea otters are carnivorous (meat eaters) and they consume all different kinds of ocean creatures including sea urchins, shellfish, fish, and various marine invertebrates. Each individual otter has its own favorite foods, too. One may prefer sea urchins; while another prefers abalone.

A familiar position for the sea otter, which eats, rests, and sleeps while lying on its back. *NOAA/Commander John Bortniak, NOAA Corps (ret.)*

These animals spend a considerable amount of time looking and diving for food because they must eat constantly to survive—they consume the equivalent of 20 to 25 percent of their body weight each day. They often use their front paws (forepaws) to dig into the seafloor to find burrowing animals such as clams. They are also one of the few animals to use "tools" to get to their food. For example, when an otter dives for abalone, it may use a stone to "hammer" the shellfish if it is stubbornly stuck to a rock. It may also use a stone to dig for clams. The sea otter has a small pouch under its left armpit, where it can store a rock or some food while swimming.

Once the sea otter resurfaces with its catch, it eats—or prepares to eat. These animals have strong teeth that can bite through crab shells and a few other types of shellfish. But hard-shelled fish such as abalones must be cracked open; some sea otters accomplish this by hitting the shell with a rock, as the shell sits on the animal's chest, while others will put the rock on their chests, then smash the shell against the rock. Even if a sea otter doesn't use rocks as tools, as is the case with the Alaskan sea otter, which has plentiful soft- and thin-shelled animals to prey on, they still have the ability to do so.

How does a **sea otter rest**?

Sea otters seem to do everything on their backs—and they do. They eat, rest, and sleep while lying on their backs. When an otter sleeps, it will generally wrap itself in kelp to keep from drifting away. These animals lie in the waters with all four limbs—which are not protected by fur—projecting out of the water. They may do this to conserve heat; unlike other marine mammals, the sea otter does not have a layer of insulating blubber. Instead, their fine, dense fur—the densest of all the animals, with up to 1 million hairs per square inch (15,000 per square centimeter)— traps air and helps the animals maintain warmth.

Have any **sea otters** become **extinct**?

At one time, scientists thought the southern sea otter, found off the shore of California had become extinct—mainly because of the fur trade during the nineteenth century. But now we know a small group survived around Big Sur, a group responsible for eventually repopulating the otters along the California coast. Today, there are about 2,200 sea otters along a 250-mile (402-kilometer) range along the state's central coast; the original population was about 20,000. And there are still some concerns: A survey during the late-1990s indicated there was a drop in the southern sea otter population over the previous few years. Scientists are not sure if the decrease is due to disease, contaminants, or to accidental drowning in fish traps.

What happened to **sea otters** along **Alaska's Aleutian Islands**?

The population of sea otters in certain habitat "pockets" along Alaska's Aleutian Islands has plummeted by 90 percent in less than a decade. Scientists were not sure why until just recently: They believe that hungry orcas have been eating the sea otters like popcorn. The problem extends even further: Without the sea otters, the population of sea urchins (which the otters normally eat) is booming, stripping the undersea kelp forest.

Why have the orcas (killer whales) developed an appetite for sea otters—which they usually ignore while hunting sea lions and harbor seals? Scientists believe that over-fishing and rising water temperatures may be the problems, as both have caused a decline in the fish populations. Commercial fleets over-fished the area; then, in the late 1980s, the populations of sea lions and harbor seals—animals that depend on the fish for survival—declined to 10 percent of what they once were. With fewer sea lions and harbor seals to prey on, the orcas turned to the sea otters as a food source; the whales moved from the deeper, open ocean to the coastlines in search of them. The problem in the Aleutian Islands is a clear example of what happens when just one link in the food chain is disturbed.

MARINE BIRDS

What is a **bird**?

A bird is a warm-blooded vertebrate that reproduces by laying eggs; they are thought to have evolved from reptiles. Birds are members of the animal kingdom, and have their own class, the Aves. They have four limbs, with the front two limbs modified into wings.

How many **species of birds** are found along the **coasts** and **oceans**?

There are only about 300 species of birds that live on the coasts and oceans, which is equal to about 3 percent of the total number of birds in the world (on land and water). What they lack in diversity, they make up in their huge numbers. Most of the bird concentrations occur where sea life is prolific, such as in the Antarctic Ocean (also called the Southern Ocean), off the coasts of Peru and Chile. Birds that live off the ocean are usually divided into two groups, seabirds and shorebirds. The diet of these birds is varied and can include worms, crustaceans, bivalves, fish, snakes, and mice.

What are **seabirds**?

The name seabird, also sometimes called oceanic bird, is used by some ornithologists (zoologists who study birds) to describe any bird that spends much of its life on or over the ocean waters. These birds include some familiar water bird species, such as boobies, gannets, frigate birds, loons, cormorants, puffins, pelicans, certain ducks, gulls, and terns.

"Seabird" especially applies to albatrosses, shearwaters, petrels, and storm-petrels—those birds that are seen most often over the ocean waters. Except during breeding season—or when they are seen following ships for wastes thrown overboard—these birds live far from land. Amazingly, the seabirds are some of the most abundant birds that fly over the oceans. In fact, one of them, the Wilson's storm-petrel, was once referred to as the most abundant bird in the world.

Red-crested cormorants and distinctive horned puffins (with their black topcoats and compressed beaks) share a perch atop a rocky precipice. *NOAA/Captain Budd Christman, NOAA Corps*

Does any **bird** spend its **entire life at sea**?

Yes, the albatross spends almost its entire life on the ocean, only coming to land every other year to nest.

What are **shorebirds**?

Shorebirds are the many species of birds that wade into shallow waters—of estuaries, rocky shores, beaches, tidal flats, jetties, and marshes—to find their prey, but do not swim. These birds seem quite similar to each other, especially the sandpipers. Shorebirds are usually "counter-shaded," or dark on top where they receive the most sunlight, and light on the bottom. This makes the birds hard to see on a beach—and even more difficult to determine their species. Thus, bird watchers and ornithologists identify the animals by their beaks, legs, or behavior.

Do **shorebird and seabird bills** differ from one another?

Yes, just as with land birds, they differ greatly, and there is even a variety of bills within each group. Most birds have bills that are specially adapted so they can hunt and eat food in whatever environment they live in—on land or in the ocean. For example, the red-breasted merganser, a shorebird, has a spike-like bill with saw-tooth margins, which it uses to grab and tear open crustaceans. The greater yellowlegs, another shorebird, has a long, thin beak it uses to stab or grab its fish or crustacean prey, or to probe in the mud. The bill of the albatross, a seabird, is specialized for catching fish in the open ocean.

An osprey guards its nest high atop U.S. Coast Guard navigation equipment in Indian River inlet, Delaware. This shore-dweller feeds only on fish, and is often called fish hawk. It can be found along seacoasts as well as along inland lakes and rivers. *NOAA/Personnel of NOAA Ship PEIRCE*

What are some **common shorebirds**?

There are many shorebirds around the world—too many to mention in these pages. But some of the more familiar ones are the sandpipers (wading birds having long, slender bills, which they use to probe the shallow waters or muddy areas for crustaceans or worms); turnstones (short and squat birds that turn over rocks and seaweed looking for food); oystercatchers (a chicken-sized bird with a long, chisel-like bill to open up bivalves); and yellowlegs (sandpiper-like birds that prey on fish and crustaceans).

What are **wading birds**?

Wading birds are those shorebirds that walk along and in shallow-water areas; they have extremely long legs and widespread toes that keep them from sinking in the mud or sand. Most also have very long necks, which the birds fold into an S or hold straight out when they fly. Their bills can easily probe the mud or snap up (or stab) fish in shallow water. They usually nest with other shorebirds in large colonies called rookeries, making their twig-lined nests in low shrubs or empty trees in marshes and mangrove swamps.

The red-backed sandpiper, with its long beak, is a shorebird. *Field Mark*

What are some **common wading birds**?

There are many common wading birds, including egrets, herons, and ibises. Some of the more familiar ones are the snowy egret (a white-feathered bird, having a thin, black bill, black legs, and bright yellow feet); great blue heron (a 4-foot, or about 1-meter, grayish-blue bird that eats a varied diet, from mice and snakes to fish); and a white ibis (a white, heron-looking bird, but with a curved bill).

What is a **sea duck**?

Although ducks are not usually associated with the oceans, there are many species that live in bays and estuaries and along open shorelines. Most of these ducks are divers—they dive under the water to search for food, rather than feeding at the surface. Sea ducks include the red-breasted merganser (it lives mainly on fish, which it chases underwater); common eider (the largest species of duck; blue mussels are their favorite food, which they swallow whole using their stomach muscles to grind up the shells); and the oldsquaw (it feeds on fish, shrimp, and mollusks down to 200 feet [61 meters], using its wings to propel itself underwater).

How are **gulls** defined?

Gulls are very common along most any shoreline—and even inland in larger country fields. What is commonly called a seagull is actually a herring gull, seen along the East Coast and interior (near waterways) of the United States. They have long wings and slightly hooked bills and usually have a mix of white, black, and gray feathers; very rarely do they have any color. Gulls are omnivores, meaning they eat both meat and plants, and many scavenge for food; they feed on nesting sites (eating the eggs), crabs, shellfish, and almost anything that is considered food—even the sandwich of a picnicker at the beach.

What are **terns**?

Terns are shorebirds that are similar to gulls, but are usually smaller and more streamlined; they have forked tails, whereas gulls have a splay of tail feathers. Terns also hover in the air, then dive into the water to catch fish. The most prevalent is the common tern, a pigeon-sized bird with a severely forked tail. Terns are sensitive to human intervention during their mating season—thus many nesting sites have to be protected while these birds are breeding.

What are **skimmers**?

Skimmers are crow-sized shorebirds, with large, knife-like bills. One of the most common is the black skimmer—black on top, white on the bottom, and a red beak that the bird uses to skim across the water in order to catch fish and crustaceans.

What are **cormorants**?

Cormorants are black, goose-sized shorebirds that fly with their necks outstretched. They are often seen diving underwater from a swimming position, not from the air, in search of fish to eat. A cormorant uses its webbed feet and sometimes its wings, to propel itself through water, employing its tail as a rudder. Varieties include the double-crested cormorant (with an orange throat-pouch and a barely visible double-crest on its head) and the great cormorant (larger with a white patch on its throat). These birds can be an important part of their environment: They rapidly process fish into nutrient-rich droppings, or guano; algae grow in these areas, supplying food to large populations of invertebrates and fish.

What are **pelicans**?

Pelicans are large birds that feed on fish and crustaceans. This bird is best known for the pouch under its beak—a fleshy throat pouch that inflates when the pelican is underwater catching prey. This pouch is not to store the fish, but is used as a scoop to separate the fish from the water. The most common are the brown pelican (with a wing span of about 7 feet [just over 2 meters]; they also dive from the sky to catch fish such as herring, mullet, and menhaden near the surface) and the white pelican (with an amazing 9-foot [almost 3 meter] wing span; they nab fish while swimming rather than diving from the sky).

Why are **brown pelicans** a **protected** species?

Brown pelicans were close to extinction in the 1970s. Like many susceptible birds, they were decreasing in population because of exposure to DDT, a now-banned pesticide. Similar to the California condor and other bird species, their egg shells became thin, and they experienced poor reproduction rates. Thanks to the protection they received as an endangered species, they are now common along both the East and West coasts of North America, from Virginia to Florida, and from San Francisco to Mexico—and have been upgraded to a protected species.

Brown pelicans are still protected because of humans: Much of what were once prime habitats for the pelicans are now the sites of condominiums and sea walls. There is another problem: Scientists have recently discovered that near places like Miami, Florida, people who feed the birds chunks of fish left over from a day of fishing are unwittingly causing the birds a painful death. Many of the fishermen clean their catch, then feed the pelicans the leftover carcasses. But this doesn't help the birds: Much of the fish that is being fed to the birds is grouper or even dolphin—and pelicans cannot digest the bones of these larger fish. After consuming these handouts, the pelicans fly away, and the bones either get caught in their throats or press against their stomach lining, puncturing the stomach or other organs. In Florida, marine agents are now posting educational signs at marinas statewide, alerting fishermen to the dangers of feeding the pelicans. If people still want to feed the birds, marine agents recommend giving them boneless chunks or the smaller fish that are part of their natural diet—such as grunts, mullet, or pinfish.

Do any birds fly **over the oceans** when they **migrate**?

Yes, many of the migration paths of various birds—seabirds and land birds (mostly the songbirds)—pass over the oceans. For example, a seabird called the Leach's storm petrel breeds on rocky islands and deserted coasts from Canada to Massachusetts in the spring, then migrates to the open ocean during the winter. The inland-dwelling ruby-throated hummingbirds migrate south from the United States every fall, covering 600 miles (965 kilometers) including the Gulf of Mexico, to follow the seasonal growth of flowers (besides beetles, bugs, flies, gnats, and other insects, hummingbirds also eat the nectar of flowers). In the spring, they migrate in the opposite direction.

The penguin may not be able to fly but it sure can swim. Here, a pair of Gentoo penguins ably navigate the waters. *CORBIS/George Lepp*

How did certain **ocean birds** become **flightless**?

Some ocean birds became flightless because in their environments, mostly oceanic islands, they had no natural predators—so they no longer needed their wings for escape. This group may include the flightless penguins in the Antarctic, as well as the flightless grebes, rails, and cormorants that live, or have lived, on oceanic islands. Many of these flightless birds have also become extinct. One of the most famous was the dodo, an ocean island bird hunted to extinction by humans.

Which **birds** are the **best swimmers**?

Among birds, penguins get the award for best swimmers. They are flightless marine birds but their streamlined shape makes them excellent swimmers. They can stay in the water for a long time without being affected by the cold: They carry about a 1-inch- (2- to 3-centimeter-) layer of fat and have waterproof feathers, with a layer of air trapped underneath for added protection from the cold. Penguins can move swiftly through the water, up to 22 miles (35 kilometers) per hour, but only stay underwater for about 2 to 3 minutes at a time. They are so fast that they can shoot out of the waves and land securely on the ice—upright on both feet. And they are agile and quick on land, too: They walk upright, and in snow, they often move by sliding on their bellies.

How many **types of penguins** are there?

There are many types of penguins around the world, mostly found along the coasts of the Antarctic, in the southern temperate cool zone, on the Galapagos Islands, and on the subtropical coasts of South America, South Africa, and Australia. Among the most well known penguins are the emperor and king (these two penguins are the largest and they are members of a group called the Giant penguins); other groups are the Adelie, crested, black-footed, yellow-eyed (or yellow-crowned), and dwarf—all of which include many species. They feed on planktonic animals, especially krill, and small fish, crabs, and squids.

MARINE REPTILES

What are **marine reptiles**?

Marine reptiles are very similar to terrestrial reptiles: They are cold-blooded, air-breathing animals covered with many scales. And although 14 percent of the 40,000 known species of vertebrates are reptiles, there are very few marine reptiles. They include the crocodile, sea turtle, and sea snakes. Most of these animals are still tied to the land, as they have to lay their eggs on the shore; when the eggs hatch, the hatchlings resemble miniature versions of the adults.

What is a **crocodile**?

Crocodiles are found in warmer climates worldwide. They have long bodies and powerful tails used for swimming or defense. In water, they use their limbs to maneuver and steer; on land, they use them to walk with a slow gait, with their bellies held high off the ground. Crocodiles are usually fairly aggressive and may bite or swing their tail in defense. These animals eat a wide variety of food, especially fish, large vertebrates, and carrion.

One of the more well-known marine crocodiles is the American crocodile, which lives exclusively in the saltwater areas of southern Florida and the Florida Keys. Because of years of over-hunting (their hides were highly valued), they are now an endangered species. But this designation has not protected them enough; their habitats are still being encroached upon (due to development) and they are the targets of poaching (illegal hunting).

A yellow-bellied sea snake, a marine reptile, slithers along a South African beach. *CORBIS/Anthony Bannister; ABPL*

What is the **difference** between a **crocodile** and an **alligator**?

There are two ways to tell these animals apart: by their size and their head. Crocodiles (*see* photo, next page) are slightly smaller and less bulky than alligators, but the crocodile's head is larger and it has a narrower snout and a pair of enlarged teeth in the lower jaw that fit into a notch on each side of the snout; the upper and lower teeth are visible when the jaw is closed. Alligators have broader bodies and snouts; when this animal's jaw is closed relatively few upper teeth are visible.

The American alligator and the American crocodile are also differentiated by their habitats: the alligators live in fresh to brackish (slightly salty) water; the crocodiles live in salty water.

It would be a menace at half the size: This huge crocodile was caught in 1925 off Borneo, but the marine reptile can be found in warm climates worldwide. *NOAA/Family of Captain Jack Sammons, C&GS*

What is a **sea turtle**?

A sea turtle is a reptile with a lightweight, streamlined shell, which protects the turtle's vital organs. It has a heavy neck that, unlike many land turtles, it cannot pull all the way into its shell. Its legs are muscular and powerful, allowing some species to swim at speeds of up to 35 miles (56 kilometers) per hour. There are 8 species of sea turtle, including the green, loggerhead, hawksbill, and leatherback. Sea turtles range widely in size: the olive ridley grows to less than 100 pounds (45 kilograms), whereas the leatherback can be anywhere from 650 to 1,300 pounds (295 to 590 kilograms) at maturity! Certain species can also live very long—more than 100 years.

What **problems** have plagued **sea turtles in Florida**?

Most sea turtle species in Florida are either endangered or threatened. One of the more recent problems they face is papillomas, a potentially fatal disease that causes tumor-like growth on the soft tissue of sea turtles in this region. The growths can cover the eyes of the turtles, causing blindness and eventual death, as the animal cannot see to find food. The actual cause of the papillomas is unknown, although several researchers believe it may be caused by waters that are polluted by runoff.

Another problem endangering the sea turtle is the rise in boat-related injuries—something that may increase during high nesting years (when there are a number of nests on the beaches). During this time, there are more turtles—and, usually, more people on the water. Though the turtles aren't usually at the surface, where they might be in danger of being hit by a boat, they do surface for fresh air; the more often they do so, the greater the likelihood for a mishap.

Sea turtles have muscular and powerful legs, allowing some species to swim at speeds of up to 35 miles (56 kilometers) per hour. *NOAA/OAR National Undersea Research Program; G. McFall*

Loggerhead sea turtles, hatched on a Virginia beach, head for the ocean. ***CORBIS/Lynda Richardson***

How are **endangered sea turtles** being protected?

Many people are trying to find ways to protect endangered sea turtles. For example, in Indian River County, Florida, officials have developed a habitat conservation plan as part of a legal settlement. When the county wanted to put up sea walls, the Sea Turtle Survival League brought a lawsuit, citing that such walls would interfere with nesting habits: They argued that the proposed walls would not allow the turtles to dig nests in the area, and would also lead to the erosion of turtle-nesting sites farther down the beach. The main reason for the concern was that this area, along Florida's Atlantic coast (about two-thirds of the way down the Florida peninsula) hosts the world's second-largest population of threatened loggerhead turtles—and almost all the nesting sites of endangered green and leatherback turtles. This ranks Indian River County as a globally important sea turtle nesting area.

What is a **sea snake**?

A sea snake is similar to a land snake, except it spends its time in the water. These reptiles are found in shallow tropical and subtropical waters (but not in the Atlantic Ocean)—sometimes numbering in the hundreds. The sea snakes are all related to the cobra family, and have a very potent venom (it can cause severe injury to humans). They use their flattened tails as a paddle as they swim, and they usually feed on fish. Between breaths at the surface, they can stay under water for more than 30 minutes. An example of this animal is the yellow-bellied sea snake.

FISH & OTHER OCEAN LIFE

DEFINING FISH

What is a **fish**?

A fish is an aquatic animal—a vertebrate (skeletal) creature that lives the majority of its life underwater. But there is such a wide variety of fish—from the flying fish that live just above the water's surface to those that are found deep in the ocean trenches—that neatly defining them is difficult. Still, some generalities can be drawn: Most fish are cold-blooded, breathe by means of gills, and have a heart with only two chambers (as compared with the four-chamber heart of humans); others are not cold-blooded and do not have gills. Some have scales, others have rough skin composed of tiny "teeth" called denticles. Most have fins for swimming, but they come in various combinations, sizes, and shapes.

The reasons for such diversity are these: In adapting to their diverse environments, fish have made many evolutionary changes over hundreds of millions of years. For example, there are blind fish, fish that can crawl about on land, and electric fish; others build nests, change color in a few seconds, or even puff up to ward off predators.

When did **fish evolve**?

It is difficult to say when fish actually evolved, but scientists do know they are the descendants of jawless fish. The following lists the possible evolution of the fish over time:

About 460 to 480 million years ago: The first jawless fish evolve. (It's worth noting here that this starting point is hotly debated in the scientific community.)

About 450 million years ago: The first jawed fish emerges.

About 390 million years ago: The ancestors of the bony fish evolve.

About 380 million years ago: The first shark-like fish evolves.

About 360 million years ago: The early offshoots of the bony fish (Osteichthyes) evolve into the first amphibians (intermediates between fish and reptiles).

About 175 million years ago: The first true bony fish emerge.

Between 190 and 135 million years ago: The first modern sharks evolve.

Why were **fish important** to the **evolution of life** on Earth?

Fish were very important to the evolution of life on Earth because they were the precursors to amphibians, the first animals to venture onto land.

How did **amphibians evolve** from **fish**?

Amphibians—an offshoot of fish and the first creatures to crawl on land—adapted to their new terrestrial environment by developing primitive lungs and using their fins for crawling. Still, the earliest amphibians spent most of their time in the water. They were very fish-like, and laid fish-like eggs in a moist place, usually in the water.

What is **ichthyology**?

Ichthyology is the study of fish. Swedish naturalist Peter Artedi (1705–35) is considered the father of ichthyology; he made several investigative writings about fish. His knowledge of the relationships

Salamanders, such as the Cheat Mountain salamander shown here, are among the vanishing species of amphibians—living links in the evolutionary chain. This species survives only in West Virginia. *Associated Press/Marshall University File*

among the various species and his concepts of fish groups have become classics in the field.

What does it mean to be **cold-blooded**?

When it comes to ocean fish, cold-blooded means that the body temperature is just about the same as the surrounding water. This means that fish are particular about the temperature of the water in which they live. In fact, most species are sensitive to temperature changes, and usually cannot stand rapid fluctuations of more than 12° to 15° F (7° to 8° C).

But not all marine fish are cold-blooded. As far as we know, certain tunas and bonitos are exceptions to the rule. For example, the blue-fin tuna can regulate its own body temperature, similar to a mammal.

What is the **smallest known fish**?

So far, the *Pandaka pygmaea* of the Philippines is the smallest known fish. It measures just under a half inch (1 centimeter) at maturity.

What is the **largest known fish**?

The whale shark is the largest known fish, measuring up to 60 feet (18 meters) in length—and amazingly, this giant only feeds on plankton. The whale shark can weigh up to 15 tons—or 5 billion times as much as the common goby.

Where do most **ocean fish live**?

Most ocean fish live along coastal areas, especially on the continental shelves to depths of 400 to 600 feet (122 to 183 meters). These coastal fish occupy almost every niche (habitat in this area), and represent two-thirds of all known living fish species, with most of the fish living within 20 miles (32 kilometers) of the shore. There are fish (such as flounder, sole, and cod) that blend in with the bottom sands in the shallow coastal waters; others spend their lives in the ocean, but travel upstream through inland rivers to spawn, such as salmon and shad; still others occasionally walk on land, such as the walking catfish.

Fish are also prolific in the open ocean waters, at depths from the surface to 6,000 feet (1,828 meters). Fish such as tuna, mackerel, and marlin prefer waters no deeper than 3,000 feet (914 meters)—a layer of water where temperatures are most comfortable for them.

The fewest number of fish live at the deepest ocean depths, past about 18,000 feet (5,486 meters). We may be surprised in the future as more ocean depths are explored: Some deep-water areas may have more fish than we realize.

Have any **new ocean fish** been **discovered**?

Many times when scientists explore new parts of the oceans, they discover new types of fish. For example, in 1978 a research group lowered baited traps and a camera through the Ross Ice Shelf and into the Antarctic waters, to a depth of 1,960 feet (597 meters) below sea level. There, they found many new types of fish—adding to the already huge list of fish species.

What is a **living fossil**?

A living fossil is any organism that was once thought to be extinct, but was actually found, very surprisingly, to still be alive. Because the discovery of such creatures provides scientists with new information about evolution, the organisms become part of the fossil record. Perhaps the most famous example of the discovery of a living fossil occurred in 1938: A small trawler (fishing boat) brought up an amazing catch from deep in the Indian Ocean, near the Comoros Islands off Madagascar—a coelacanth (SEE-la-kanth), a fish that was thought to have gone extinct millions of years ago. Other coelacanths have been found since. The surprising discovery caused scientists to revise the thinking about the life that exists in the deep (as of yet, mostly unexplored) ocean: It's possible that there are all kinds of ancient-looking fish living in the depths.

The coelacanth first appeared during the Devonian period of the Paleozoic era, almost 400 million years ago. Some scientists believe this species of fish was the first to haul itself out of ancient oceans and onto land—and are probably the ancestors of many land animals. But based on the fossil record, it appeared that the fish went extinct toward the end of the Cretaceous period, about 60 million years ago. But evidently, enough coelacanths made it through millions of years to maintain a population—and now live deep in the oceans.

What **missing fish** was recently rediscovered?

A giant roughy, also known as a giant sawbelly (*Hoplostethus gigas*), was first recorded and last seen in 1914. Now this fish has been found again after 85 years—it was discovered during a survey of commercial catches from the Great Australian Bight.

The fish was caught in water between 590 and 1,148 feet (180 and 350 meters) deep, in the same general location as the one found in 1914. Its decades-long absence is attributed to its habitat: The giant roughy apparently lives in the deep, central part of the Great Australian Bight, which is lightly fished, or not fished at all. In addition, it probably lives on or close to the rough bottom, avoiding the nets of trawlers.

The impetus for this discovery was the creation of the *Australian Seafood Handbook*—an identification guide for edible Australian

How does a fish breathe?

A fish breathes oxygen present in water: The water enters the fish's mouth and passes through a chamber over the gills; as the oxygen-rich water flows over the gills, a series of membranes allow air, but not water, to pass through; the oxygen then passes into the bloodstream; finally, carbon dioxide is released into the water as a waste product.

This system is more efficient than the one used by those of us on land: Most land mammals absorb only about 20 percent (or less) of the oxygen from the air into their bloodstreams, while fish absorb up to 80 percent of the oxygen available in the water.

fish. Scientists had been working closely with commercial fishermen to sample and record catches from the deep oceans and coasts. While recording and photographing commercial species that are caught off the Great Australian Bight, a scientist was shown a selection of fish caught by a trawler. One of the fish was the long-missing giant roughy.

The discovery of this long-lost fish provides yet another example of how elusive the life forms in the deep ocean can be.

How **long** can a **fish live**?

Although not all fish species have been observed over time, it is thought that most fish live an average of about 25 years. Scientists do know the average age (or life span) of those fish that have been extensively studied or monitored; for example, Atlantic cod and herring live about 22 years, haddock and barracuda live about 15 years, and a bluefin tuna can live about 13 years. But some fish remain to be studied, so the average life span estimate may change as more research becomes available.

How do **ocean fish stay afloat**?

Most ocean fish stay afloat with the use of a gas, or air, bladder. This gland-lined airtight sac is located in the gut of the fish; the size dependent on the size of the fish. The glands extract gases from the fish's bloodstream, and direct these to the bladder. The fish has the ability to regulate the amount of gas in the bladder, allowing the fish to travel up and down very rapidly in the water column—but the bladder usually only works down to about 1,200 feet (360 meters). In fact, hooking a fish and bringing it to the surface too fast can cause the fish's bladder to burst—causing its stomach to be forced out of its mouth.

Some fish, such as mackerels, do not have gas bladders. These fish have to swim and move constantly to maintain a certain level in the water—if they don't, they will sink.

How do **fish swim**?

Most fish swim by pushing water aside—by wiggling the head and tail and sometimes the fins, depending on the fish. The fins are used for steering and stabilization: The pectoral (side) fins are used for balance or turning, and for pitch (tendency to ascend or descend); the dorsal (top) and pelvic (bottom) fins are used as stabilizers, controlling the body's rolling motion; and tail fins are multi-functional, and are used for propulsion, steering, and stabilizing. Most fish also reduce their resistance to water with a coat of slimy mucus, permitting water to slip freely over their bodies.

There are exceptions, of course. For example, flatfish, because their bodies are flattened sideways, undulate their bodies from side to side to move. Flat rays and skates also undulate their bodies, and move their pectoral fins up and down instead of side to side.

How **fast can fish swim**?

Amazingly, there are some fish that swim extremely fast—moving faster through the water than many of the speediest terrestrial animals can move across the land. Some of the fastest swimmers are the sailfish and swordfish (one unconfirmed report clocked a swordfish at 150 miles

What is a school of fish?

A school is a group in which all the individual fish are headed in the same direction. Like a flock of starlings in the air, the fish are uniformly spaced and swimming at the same speeds. It is no doubt an evolutionary feature in many species of fish: As the group moves in unison, it is more difficult for a predator to catch its prey. Plus, because there is such a large group, it is easier to confuse a predator—or even fool the predator into thinking the "creature" (the school) in its path is huge. Changes in the behavior of a school of fish depend on the surroundings, with the school responding to all sorts of variations in the environment, including noise, sudden motion, or a strange presence (such as a human swimmer or wader).

[241 kilometers] per hour). The following list gives the approximate recorded speeds of certain ocean and, for the sake of comparison, freshwater fish (indicated by an asterisk, *).

Fish	Speed distance per hour
sailfish	68 miles (109 kilometers)
swordfish	60 miles (97 kilometers)
blue-fin tuna	50 miles (80 kilometers)
shark	22.4 miles (36 kilometers)
trout*	21.7 miles (35 kilometers)
pike*	15.5 miles (25 kilometers)
carp*	7.5 miles (12 kilometers)

Can **fish fly**?

Not really, although there is something called a flying fish. But they do not fly—they glide. Propelled by the movement of its strong tail, this fish leaps into the air at speeds of up to 20 miles (32 kilometers) per hour. It uses its wide pectoral (side) fins as wings, gliding close to the ocean surface. By flicking its tail against the water, it is often able to get an additional push for a longer "flight." Some flying fish have been known to soar as high as 20 feet (6 meters) and glide for 1,300 feet (396 meters) along the surface.

The duckbill eel, like other eels, is a bony fish (order Osteichthyes) that, surprisingly, has an acute sense of smell. This one was found off the coast of Hawaii, at a depth of about 2,600 feet (780 meters). *NOAA/OAR National Undersea Research Program*

Can **fish walk on land**?

Yes, there are some fish that walk on land. Walking catfish can survive long periods on land and are capable of breathing air, using a lung-like organ as a supplement to their gills. The lungfish can also walk on land, using the swim (or gas) bladder as an air reservoir, and a nose-to-mouth connection that allows it to breathe the air.

Can **fish smell**?

Yes, fish do have a sense of smell. The nostrils of most fish are located in the front, just behind the mouth opening. These organs are so powerful that some fish can detect the secretions from a predator's skin long before the animal gets close.

In fact, some fish don't depend on their sight as much as they do their sense of smell. Sharks and rays depend on smell to detect prey; eels have the most acute sense of smell, using long nasal sacs to detect a scent. Certain species migrate across the oceans to spawn by detecting very minute changes in seawater chemical variations—which they pick up by smell.

How do **fish see** underwater?

In most cases, because fish have to see in low light, they have no eyelids and no irises; (in the human eye, the pupil responds to the amount of light available, opening or closing as needed). These characteristics allow the fish's eyes to gather the most light—which also gives the animals what is often described as a "bug-eyed" appearance. Fish have monocular vision—the eye on one side of the head is sensed by the opposite side of the brain. A fish's eyes are located on each side of the head, allowing the fish to see in any one direction at a time. Many also have upward-directed eyes; during daylight, they detect prey by spotting its silhouette against the sunlight from above. And of course, there are exceptions: a few surface-dwelling fish have eyes adapted to see in both water and air.

Although scientists believe that most fish have some color vision (except sharks and rays, which may only see in black and white), no one really knows if the animals are affected by color. Are they actually coming after the pattern of a fisherman's hook or the color(s)?

Some believe a fish's pattern may be most important to other fish. For example, adult and young angelfish are brilliantly colored and patterned—but they look completely different from each other. One theory is that the adult recognizes the young's patterns among the huge variety of fish found in a reef—and thus they will not threaten the young of their species.

Can **fish hear**?

Yes, fish can hear. Sound is as important to underwater communication as it is to communication on land. Fish, like other organisms in the sea, use sounds to attract mates and to detect prey or predators.

Fish hear much the same way as humans, although they lack the air-filled middle ear—in other words, there are no ossicles like those found in a mammal's ear. Fish also do not have a cochlea, but do have hair-bearing cells in a chamber-like organ called the lagena. Most fish have three semicircular canals—although lampreys have two and hagfish only one.

Are some **fish "electric"**?

Yes, some fish use electric fields to detect prey, navigate, or even to defend themselves or attack other fish. These fish, including skates and rays, have organic "batteries" that evolved from muscles or nerves. Electric rays found in sandy or muddy shallows or in moderately deep seas can generate more than 200 volts of energy. After a number of shocks, mostly to immobilize prey, its "batteries" are completely depleted, and the animal takes several days to recharge. Electric eels, which are not true eels, can generate up to 600-volt pulses from "batteries" that extend almost half the length of their bodies—which can reach up to 8 feet (2.5 meters) long.

Why are **fish important** to **humans**?

Fish have been, and continue to be, one of the most important commercial products of many countries. For years, various species of fish (such as tuna, herring, and sardines) and certain "industrial fish" (monkfish, sea robin, squirrel hake, shark, and ray) have been used in livestock feeding (the fish are an excellent source of nutrients and oils). Others have been used as major food sources for humans, especially for populations living on islands or along the coasts, where fish are, in most cases, more cost-effective to harvest and eat than are other types of animals.

Are **fish indicators** of **environmental change**?

Yes, fish—especially their demise—can be indicators of environmental change. Plankton, the tiny mobile organisms in the upper levels of the sea, are at the bottom of a huge food chain (or food pyramid), supplying a major part of the dietary requirements of many fish. The progression of eating in the oceanic food chain is from tiny to huge (fish are in the middle of this chain) and keeps the ocean ecosystem in balance. If the plankton population is affected by such problems as environmental pollution, ozone hole depletion, or temperature changes, fish will suffer, too. The local food chain can then break down—and in turn, may eventually affect the global food chain or food web.

CLASSIFYING FISH

How are **fish classified**?

Fish are classified in the kingdom Animalia, phylum Chordata, and class Pisces (although in some classifications Pisces is listed as a subclass). In the Pisces class (or subclass), 90 percent of the fish are Osteichthyes (bony fish), which accounts for some 30,000 different species. The remaining 10 percent are Chondrichthyes (cartilaginous fish), with 625 species, and Agnatha (jawless fish), with 50 species. Nearly half of the 40,000 or so known species of vertebrates ("segmented backbone" animals) are fish. (All together, there are seven classes of vertebrates: Agnatha, Chondrichthyes, Osteichthyes, amphibians, reptiles, birds, and mammals.)

But this classification system, though widely accepted, is not standard; there is still disagreement over the classification of fish, which means there are alternate systems of organization. For example, one classification breaks down classes of fish into Agnatha, Placodermi (known only as fossils), Chondrichthyes, and Osteichthyes. Further breakdown of the Osteichthyes includes the subclass Actinopterygii (ray-finned fish), superorder Teleostei (spiny-finned fish, with 95 percent of all living fish in this group), and subclass Choanichthyes (lobe-finned fish).

Do any fish have a **combination** of **bony and cartilaginous** characteristics?

Yes, even though these two major classes dominate, there are some fish that carry both characteristics. One is the coelacanth, a fish that was at one time thought to be extinct.

What are some common **characteristics** of **Osteichthyes**?

The Osteichthyes (bony fish)—including tuna, sailfish, eels, cod, and salmon—are all bony and have jaws. An early offshoot of this group may have eventually led to amphibians, the first animals to venture on to land millions of years ago (although some scientists believe the first amphibians were offshoots of a creature resembling or related to the coelacanth).

Most fish are bony fish (Osteichthyes), with skeletons made mostly or solely of bones. On each side of the bony fish, a cover protects the gills; many also have an internal organ called a swim bladder (also called a gas bladder) that helps the fish remain afloat. These fish also generally spawn rather than mate, with the males fertilizing the eggs after the females lay them. They eat a wide variety of foods, including other animals (insects, larvae, and other, smaller fish) and some plants (including phytoplankton).

Salmon, seen here in Alaska's Naknek River Estuary (at Bristol Bay) are one variety of bony fish (order Osteichthyes). *CORBIS/Natalie Fobes*

What is the **oldest bony fish** fossil found?

The oldest bony fish fossil found so far is the *Psarolepis romeri,* a fish that lived more than 400 million years ago. The fossil was uncovered in southwestern China, and has features from both the primitive and more advanced branches of the fish family. But the true ancestry of this predator is still debated—at least until scientists can find more such fish in the fossil record.

How do **flatfish differ** from most **bony fish**?

A flatfish, such as flounder or halibut, begins its life as a normally-shaped fish—complete with one eye on each side of its head. But as it matures, the fish changes: One eye moves and the mouth twists so that as an adult, it can lie on one side—making it truly a flat fish. These fish can also change color and patterns, allowing them to match the shading and texture of the seabed (this is a means of self-defense). They can also bury themselves in the sands of shallow waters to protect themselves from predators.

Is a **seahorse** a fish?

Yes, a seahorse is one of the strangest fish in the ocean. This fish actually has a head that resembles a horse; the rest of its body is long, with a curled tail at the end. It spends most of its time in a vertical position, and moves up and down in the water using its swim bladder. But it is a poor

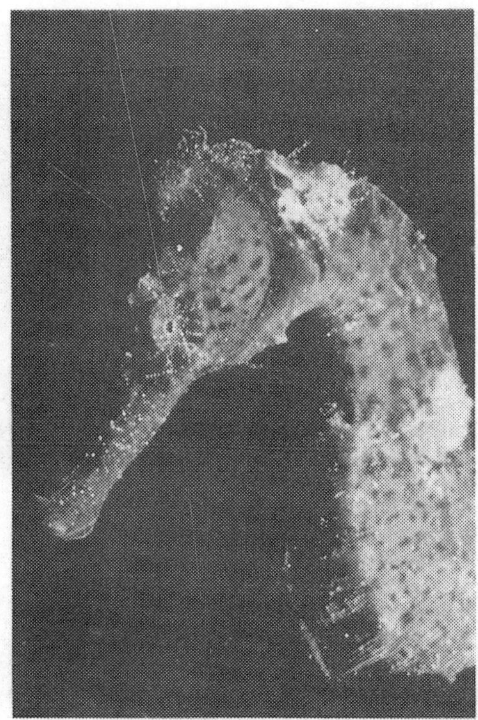

Mr. Mom: The male seahorse has a brood pouch where the female places her eggs. When they reach maturity, the offspring swim out of the pouch, looking like miniature versions of the adult. *NOAA/OAR National Undersea Research Program*

swimmer; and often uses what is called a prehensile tail—a tail that can wrap around things, such as seaweeds or sea fans—to stay in one place.

Unlike most fish, the male seahorse watches the young: the female places her eggs in the male's abdomen, in an opening called the brood pouch. When the eggs hatch, they emerge from the male as miniature versions of the adult.

What **fish** lives in **floating sargassum**?

A frogfish called a sargassumfish lives in floating sargassum algae. It rarely moves away from the sargassum, climbing over the seaweed, stalking smaller fish and other creatures that live in the algae community. Its spiny-like protrusions give the frogfish a very strange look—and its colors help it remain camouflaged in the algae. The sargassumfish can also inflate its body by swallowing air or water, preventing a predator from dragging it away as its spine catches on the sargassum.

Which **fish** is called the "**hitchhiker of the ocean**"?

Long-bodied remoras are often called the "hitchhikers of the ocean," as they take rides on sharks and other marine animals. This is called a symbiotic relationship, as the remoras eat parasites and any leftover food that surround the host. The remora clings to its host using a suction disk on the top if its head, and to release the hold, it slides forward. Certain species are highly selective about their hosts. For example, the marlinsucker and whalesuckers are so named because they prefer marlins and whales, respectively.

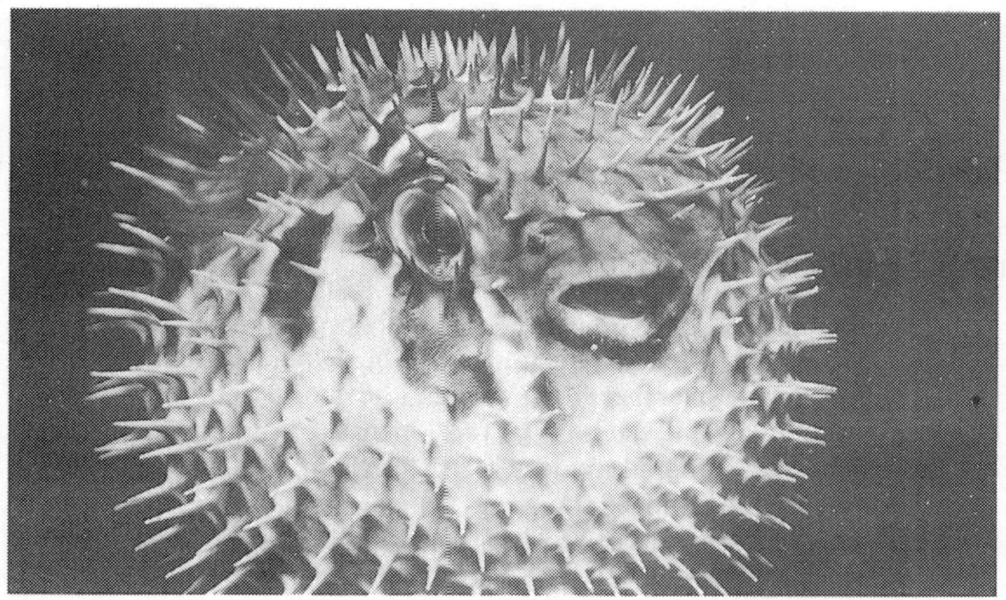

Looks can be deceiving: The spiny pufferfish can inflate itself to as much as three times its normal size to ward off would-be predators. This one was caught on film doing just that in the waters off Malaysia. *CORBIS/Jeffrey L. Rotman*

Is there a **fish** called a **dolphin**?

Yes, there is a group of surface-feeding fish called dolphins—but they don't look anything like the marine mammals of the same name. The dolphin fish has a long back fin that extends from the animal's head to its tail, and its tail fin is deeply forked. They can measure from 1 to 6 feet (0.3 to 2 meters) in length, weigh up to 75 pounds (34 kilograms), and are brilliantly colored. But they don't live long—only about 2 to 3 years.

What is a **pufferfish**?

A pufferfish (or puffer) does what its name implies: It inflates itself with water or air, enlarging itself to about two or three times it normal size. As it puffs up, it floats upside down and its spine sticks out, deterring a predator from taking a bite. When the predator leaves, the puffer deflates and flips upright again. The jaws of a puffer look like a beak and can crush hard sea urchins, mollusks, and crabs. Certain puffers are among the most poisonous marine animals known. Eating them is not advised, but some countries believe they are delicacies—and consequently, each year many people die from consuming puffers that were not prepared correctly.

What is a **goby**?

Gobies are the largest group of predominately ocean fish, with at least 800 known species—and no doubt more will be found. Most of them are very colorful, and average 1 to 3 inches (2.5 to 7.6 centimeters) in length—although some can grow up to 20 inches (51 centimeters) long. They live in many different marine niches, including mud (into which they burrow), wave-pounded shores, in sponges, or in the burrows made by crabs and shrimp. This group also includes the smallest known fish: the dwarf pygmy goby.

Is there a **fish** with **four eyes**?

Not really—but the so-called four-eyed fish does have two eyes that are essentially split in two. As the fish swims at the surface looking for insects, its eyes are divided into two by a horizontal bar. The upper part is for seeing in the air; the lower part for seeing under water. It also has the ability to see both places at the same time—or four images—as the eyes have two distinct retinal areas.

What **fish** is often called one of the "**ugliest sea creatures**"?

Some people consider the anglerfish one of the ugliest sea creatures. They are flabby and lumpy, with huge mouths filled with sharp, long, irregular teeth. They usually eat anything—and have very elastic stomachs that stretch to accommodate their prey. The anglerfish has a "lure" on the top of its head (hence its name): a long stalk with a fleshy flap at the end. This "pole and lure" wiggles around in the water right in front of the anglerfish's mouth, attracting small fish. As the small fish gets close, the angler then sucks or gulps the prey into its mouth.

The fast-swimming swordfish is considered a trophy fish. This specimen was caught by Doris Lovett Dennis, using the pole she holds in her left hand, near Santa Catalina, California, in 1936. *CORBIS/Bettmann*

Even deeper in the oceans are the deep-sea anglerfish—also considered to be very ugly. Because they live in such dark waters, the "lure" of this fish is luminous. This light is created by bioluminescence—a combination of chemicals and bacteria that glow, giving the stalk a natural light, which attracts prey.

What is a **swordfish**?

A swordfish is a fish with an elongated snout that comes to a point that measures about a third of its body length; it uses the "sword" to impale

They may not pose a serious danger to ocean swimmers and divers, but they sure look the part: Barracudas, as evidenced by the one shown (investigating a shipwreck), can look menacing indeed. ***CORBIS/Stephen Frink***

or stun its prey. Swordfish average about 4 to 10 feet (1.2 to 3 meters) in length, with some reaching 15 feet (4.6 meters); although they average less than 400 pounds (182 kilograms), they can weigh up to 1,000 pounds (454 kilograms). They are among the fastest fish in the ocean, reaching close to 60 miles (97 kilometers) per hour.

Are **barracudas dangerous**?

Barracudas are perceived to be dangerous because they have attacked swimmers and divers. But in reality, they usually only follow the diver, but don't attack. When they do attack, it is usually in sediment-filled waters: if they see a hand or foot coming their way, they think it is prey and they bite. There are about 20 species of this fish, and they average 4 feet (1.2 meters) in length. Some, such as the great barracuda (Sphyraena barracuda), can grow as long as 8 feet (2.4 meters). Barracuda are found in the Atlantic and Caribbean oceans, around the West Indies and Brazil, and north to Florida; in the Pacific, they are found from Indonesia to Hawaii.

What are some common **characteristics** of **Chondrichthyes**?

The Chondrichthyes (cartilaginous fish)—including sharks, skates, rays, and chimaeras—all have cartilaginous (fibrous versus hard) skeletons, paired fins, and jaws that evolved from gill arches. The living forms of these fish all have teeth, and are either covered with small scales or rough skin. These animals have multiple (usually five) gill slits and have essentially the same fin patterns as do bony fish (Osteichthyes). But they have different tail fins, and unlike bony fish, they have no swim bladders (or gas bladders) to keep them afloat and therefore, they must keep moving or else they will sink. They reproduce by mating rather than spawning, with most females producing live young, not eggs. Because of these characteristics, these fish can only live in certain regions of the world—primarily in saltwater.

What is **cartilage**?

The Chondrichthyes fish have cartilaginous skeletons, or skeletons made of cartilage, which is softer and more flexible than bone, but strong enough to support the animal's body. Humans also have cartilage—the "bones" in the end of your nose and in your ears are made of cartilage.

Have **shark fossils** really been found **in Montana**?

Yes, the rocks at Bear Gulch, Montana, have yielded one of the world's richest collections of fossil fish—some 113 species, including 70 species of prehistoric sharks, some of which had never been seen before. About 320 million years ago, this land was situated close to the equator and was covered by a warm, shallow sea. Here, about 60 percent of all the fish were sharks—measuring in length from a few inches (several centimeters) to about 97 feet (30 meters). Scientists believe there was a cataclysmic die-off and sudden burial of the creatures. Thus the Bear Gulch sharks are beautifully preserved, many with heart, veins, stomach contents, and skin still intact.

What is the **skin** of a **shark** like?

A shark's skin may look smooth and slippery, but it is covered with small scales that look—and feel—like tiny, sharp teeth.

Are all **sharks alike**?

No, not all sharks are alike. There are at least 370 different types of sharks, including the megamouth, sixgill, dogfish, cookie-cutter, nurse, thresher, and dwarf sharks.

What is the **largest shark**?

The largest shark—and the biggest fish in the world—is the whale shark. It can grow to more than 50 feet (15 meters) in length and weigh more than 15 tons.

What is the **smallest shark**?

The smallest shark is the *tsuranagakobitozame*—a Japanese name that means, "the dwarf shark with a long face," also called dwarf dog-shark This shark can fit in the palm of your hand; the adults measure about 5 inches (13 centimeters) in length.

What is the **least dangerous shark**?

Given their reputation (reinforced in popular media), the list of harmless sharks is surprisingly long. Most scientists agree that the least dangerous is also the largest—the whale shark. Divers have even been known to grab the back fin of this fish and take long rides through the ocean waters. These sharks feed on small fish and plankton.

Another huge but harmless shark is the basking shark, with teeth less than a half inch (1.3 centimeters) long. This shark swims slowly through a school of plankton with its mouth open, using gill rakers (a kind of strainer) to capture its food.

Even though there are numerous sharks that are thought to be relatively harmless, most scientists agree that any shark, if provoked, can use its powerful jaws and teeth to pose a real threat to whatever has provoked it.

Jaws: The great white shark, which has been noted for occasional attacks on humans, can reach more than 36 feet (11 meters) in length; its teeth can grow to an impressive 2 inches (5 centimeters). ***CORBIS***

What is the **most dangerous shark**?

Although some scientists will disagree, most point to the great white shark as the most dangerous. This fish can reach more than 36 feet (11 meters) in length, and its pointed teeth can be up to 2 inches (5 centimeters) long. But this shark is rarely seen along coastal areas, as it rarely strays into shallow waters.

Do **sharks really attack humans**?

Yes, sharks have been known to attack human swimmers and divers, but humans are normally not on the shark's menu. In most cases, a shark will try to avoid humans. Many divers note that most sharks they encounter swim away—and may be afraid of people.

But each year, about 100 people are attacked by sharks around the world, mostly by the bite-and-let-go method. It is thought that the bite-and-let-go tactic is a way for the shark to communicate to the person to leave its territory—or perhaps we humans leave a bad taste in the shark's mouth. In one study, scientists showed that many surfboards (and even people) can look like a seal to a shark swimming deep below the water's surface. And because seals are one of the favorite foods of these fish, in a case of mistaken identity, the shark will attack the human or the surfboard. Sharks have also been known to be attracted to blood in the water; and they will attack if they are bothered.

The great white shark has been noted for some attacks on humans. American author Peter Benchley wrote the famous novel *Jaws,* the story of a great white shark that repeatedly attacks swimmers off an Atlantic resort town. But other sharks, including the mako, tiger, and some hammerhead sharks, have also been cited in attacks on humans. Carpet (or wobbegong) sharks in Australia lie on the bottom of the sea during the day, their colors blending in with the surroundings. Unlucky swimmers have been known to accidentally step right on the shark, causing the fish to attack.

But there really isn't any cause for concern: The odds of being attacked by a shark are 1 in 100 million. To put this in perspective, Americans have a 1 in 66 chance of being audited by the IRS; and yet, because of the popular image of the shark, more people probably fear an attack

Though it does not prey on humans, the hammerhead shark has been cited in attacks on people. This was one of hundreds seen on a July day in 1982 off the ship *Albatross IV. NOAA/Commander John Bortniak, NOAA Corps (ret.)*

than they do an audit. And it is estimated that in any given year more people in the United States are killed by pigs than by sharks.

Are **great white sharks** really **dangerous** to humans?

New studies have shown that the danger great white sharks pose to humans is greatly exaggerated. In fact, these sharks may actually dislike the taste of humans. The great white shark, which can grow up to 20 feet (6 meters) in length with a weight of more than 5,000 pounds (2,270 kilograms), has been portrayed as a ruthless, eating machine from which no one is safe. However, although there have been 78 reported attacks along the California coast since 1926—there have been only eight deaths.

To gain a better understanding of these creatures, scientists have spent four years studying their behavior at Año Nuevo, some 23 miles (37 kilometers) off the coast of central California. These rocky outcrops are a rookery for elephant seals, the favorite prey of great white sharks.

Tagged with ultrasonic transmitters linked to a computer system, sharks were tracked night and day by a series of sonobuoys placed 1,650 feet (503 meters) apart off the western shore of Año Nuevo.

The results from this study give a much different picture of this ocean predator than the one commonly held. For example, great white sharks were found to hunt around the clock, not just during the day. These sharks were not solitary, but had social connections between pairs—and were even seen to slap their tails against the water to ward off others after a kill. Also, great white sharks were found to have picky eating habits, gently mouthing items to determine whether or not they're edible. While the soft-blubbered elephant seals feel and taste edible to the great white shark, surfboards, sea otters, buoys, and humans are not soft enough or tasty enough to eat.

What are **sixgill sharks**?

Scientists often call the sixgill sharks the "dinosaurs of the deep." There are some located off central Vancouver Island, British Columbia, Canada—in the rich waters of the Strait of Georgia. They can be up to 25 feet (8 meters) long and are often seen in water about 400 to 600 feet (122 to 183 meters) deep. They have six gill slits (thus the name), wide bodies, only one dorsal fin—and have remained virtually unchanged for millions of years.

Why is the **shark's immune system** considered **unique**?

One of the most remarkable findings about sharks concerns their immune system. In their natural habitat, sharks can ward off almost any disease, including cancer.

What are **manta rays**?

Because of the points on their fins, manta rays are sometimes called devilfish. They are the largest of the skates and rays (cartilaginous fish)—and the "flattened" relative of shark. The eyes are on the upper surface and rays have a large, wing-like pectoral fin on each side of the

The "wings" of the stingray are actually large pectoral fins that allow this fish to glide through water. This one was caught—and returned live to the waters—off the Carolina coast. *NOAA/Commander John Bortniak, NOAA Corps (ret.)*

head. The "wingspan" (distance from one end of the pectoral fin to the end of the other) of the Pacific manta can reach 25 feet (7.6 meters); the ray can weigh up to 3,500 pounds (1,600 kilograms). The Atlantic manta is slightly smaller, growing to about 22 feet (6.7 meters) and weighing about 3,000 pounds (1,300 kilograms). These graceful creatures glide through the surface waters of tropical oceans by flapping their "wings." A manta ray's mouth is small; the animal feeds on tiny fish and invertebrates it filters out of the water, often jumping completely out of the sea and diving back in to catch its prey.

What are **stingrays**?

Stingrays are often confused with manta rays, as both animals have "wings" (actually large pectoral fins on each side of its flattened body) and glide through the water. Stingrays also use their wings to excavate the seafloor, exposing shellfish, which they crush with flat, strong tooth plates. The stingray measures about 1 to 5 feet (0.3 to 1.5 meters) in diameter, its body forming a disk that includes the head and pectoral (side) fins. As the name indicates, certain types of stingrays are also known for their stinging, poisonous tail—a whip-like feature that often sports one or more sharp, barbed spines.

What is the **Agnatha**?

The Agnathans, including lampreys and hagfish, are cartilaginous fish that are jawless and have disc-shaped mouths that either grasp or suck. Unlike most fish, they have no scales and lack paired (symmetrical) fins. These vertebrates were originally filter-feeders, straining mud and water through their mouths and out their gills. Fossil remains of agnathans are evidence of the earliest vertebrates, appearing in the Ordovician period about 500 million years ago. They had bony skeletons and bony armor plates covering their bodies. Modern agnathans are greatly changed from their ancestors: over time, they have lost their bone and replaced it with cartilage.

Are **hagfish** and **lampreys parasitic**?

Yes, many of these animals are parasitic. For example, most hagfish bore into dead or dying fish, feeding until only the skin and bones remain.

Some, but not all, lampreys are parasitic, preying on other fish by sucking out their blood. For example, sea lampreys found in the Great Lakes are not native to this freshwater area; they reached the lakes after man-made canals were opened—and wreaked havoc on the fish populations until they were eventually controlled.

Do we know **all of the fish** that inhabit the **oceans**?

No, we do not know of all the fish that inhabit the oceans—far from it! In fact, scientists recently discovered four more previously unknown fish—two from the group plunderfish and two similar to jellyfish—in the frigid waters off Antarctica, near Franklin Island in the Ross Sea. This area is not known for its diversity of fish; there are only about 130 fish species there, so adding to this number was surprising.

In more detail, the newly discovered fish were:

The Antarctic gravelbeard plunderfish (*Artedidraco glareobarbatus*), which lives near sponge beds and uses a chin extension called a barbel to attract its prey (the other fish may think it's a worm). The plunderfish have an easy life—they just wait around on the seafloor, waiting for smaller fish or crustaceans to wander by, then gulp them down. They are 15 inches (38 centimeters) long; these tan-and-brown bottom-dwellers have broad, flat heads and big mouths and eyes.

The Antarctic brainbeard plunderfish (*Pogonophryne cerebropogon*), which was collected at about 1,000 feet (305 meters) below the sea surface.

Two teardrop-shaped fish were also discovered, but have not yet been named. They have jelly inside that makes them buoyant, similar to jellyfish.

How can there be **fish** in the **cold waters off Antarctica**?

Compared with the warmer waters of the world ocean, there are few fish species off Antarctica, although there are many more than you would expect. Some scientists believe that up until about 35 million years ago, a wide range of fish occupied these areas, probably because the water was much warmer then. Over time the currents and land masses changed so that today cold waters are brought to the area—but the fish remained and adapted to the changes in the local water.

Over the past 15 million years, many of these fish have flourished. They also have some specific, distinguishing traits when compared with other types of fish in warmer waters. For example, certain Antarctic fish, including the plunderfish, don't have swim bladders—so they don't rely on air for buoyancy; some have increased their body fat and decreased the density of their skeletons, which increases their ability to float. Many Antarctic fish are also thought to produce an "antifreeze" protein, which protects them from the cold waters. Another reason for their success is the lack of competition: Relatively few fish species venture into such frigid waters. This is referred to as adaptive radiation, in which a species evolves to fill niches (habitats) where other species do not venture—thus there is no competition. The lack of competition can also jump-start the evolutionary process, which may be happening to the fish in the chilly waters off Antarctica. In fact, some researchers estimate that the new species recently found off Antarctica have only been there for a few million years.

OTHER MARINE CREATURES

What **other marine creatures are there**—besides mammals, fish, birds, and reptiles?

The phyla (groups) of marine animals that have not been mentioned so far are:

Marine Worms

Porifera (sponges)—some 10,000 species

Coelenterata or Cnidaria (polyps, jellyfish, corals, sea anemones)

Ctentophora (comb jellies)

Ectoprocta (bryozoans)

Brachiopoda (lampshells)

Mollusca (this phylum is being revised, but it traditionally includes snails, conches, abalones, clams, mussels, scallops, oysters, conches, nautiluses, squid, and octopuses)—some 80,000-100,000 species

Arthropoda (horseshoe crabs, sea spiders, lobsters, crabs)—some 30,000 marine species

Echinodermata (sea lilies, starfish, brittle stars, sea urchins, sea cucumbers, sand dollars)

Tunicata (sea squirts, salps)

Arcania (lancelets)

What are **marine worms**?

Marine worm is a loose term to describe invertebrates that are either independent or parasitic. There are about 9 phyla (groups) that contain marine worms:

Acanthocephala: These animals are spiny-headed, sausage-shaped, parasitic worms. When they are eaten by a fish, they attach themselves to the fish's gut, and can often impair the digestion of the host. If the fish is eaten by another animal, such as a bird or seal, these worms can become a parasite in the new host.

Echiuroidea: These are spoon worms found around the world, and range from 0.4 to 24 inches (1 to 60 centimeters) in length. They have a strange folded organ (internal), which is used for food gathering: the proboscis may be 30 to 40 inches (90 to 100 centimeters) long—although the worm's body averages only about 3 inches (8 centimeters).

Platyhelminthes: This phylum includes 25,000 species, mostly parasitic flatworms such as flukes and tapeworms. Tapeworms may have fish as hosts; they are also known to infest humans.

Nemertea: The nemerteans are called ribbon worms, with about 800 species worldwide. The ribbon worm can grow to as much as 100 feet (30 meters) in length, and is found in both the open ocean waters and on the seafloor.

Aschelminthes: These are round worms, although this group has been questioned: All the classes within the phylum may be actual phyla themselves. Most of these worms are small scavengers that feed on dead organisms.

Phoronida: These are the horseshoe worms, and are usually less than 0.4 inches (1 centimeter) in length. They live in a membrane tube on the bottom of shallow-water areas. They can bore into corals, rocks, or shells, using an acid secretion that breaks down and dissolves calcium carbonate.

Sipuncula: These are peanut worms, and include 325 species.

Chaetognatha: This phylum is the arrow worms, and include 150 species. They are usually found in shallow water, and, when small, have also been classed as plankton—making them an important link in the continental shelf food chain. The worms average about 0.4 to 1.2 inches (1 to 3 centimeters), but can grow to more than 4 inches (10 centimeters) long. They have several fish-like characteristics, including numerous chitinous (fibrous) teeth, external swim fins, and a distinct tail.

Marine life includes tiny worms, such as the Sabellid shown here, climbing out of a tube at the University of Hawaii Institute of Marine Biology. *NOAA*

Annelida: This phylum is the segmented worms, such as marine segmented worms, leeches, and earthworms. The list of annelidans is long, and the phylum is divided into two large classes, the Polychaete and Clitellata (or Oligochaeta and Hirudinea). The Polychaete are mostly marine organisms and are highly diversified. Many move around the ocean using parapodia (appendages resembling feet), and are filter feeders, swimmers, or burrowers.

What are the **longest known worms**?

The ribbon worms are among the longest such creatures in the world. For example, a form of ribbon worm called the bootlace worm, living off the coast of Great Britain, measures an average of 15 feet (4.6 meters)—but some as long as 100 feet (30 meters) have been reported.

What is a **sponge**?

Sponges are not the synthetic kind you find in the kitchen. Filter-feeding sea sponges are members of the Porifera phylum, with more than 10,000 species in existence—all aquatic and mostly ocean-dwellers. Sponges live from the intertidal zone to depths of 26,000 feet (8,500 meters) below sea level. These simple life forms have no organs, and are not mobile. They have outer and inner layers of cells, with the inner layers grabbing microscopic food particles from the passing water currents. The currents also carry oxygen to the animal and take away carbon dioxide wastes. Certain species have chemical defenses, producing offensive tastes and smells to deter predators. Still other sponge species are parasitic, burrowing into corals and shelled animals, killing the host in order to survive.

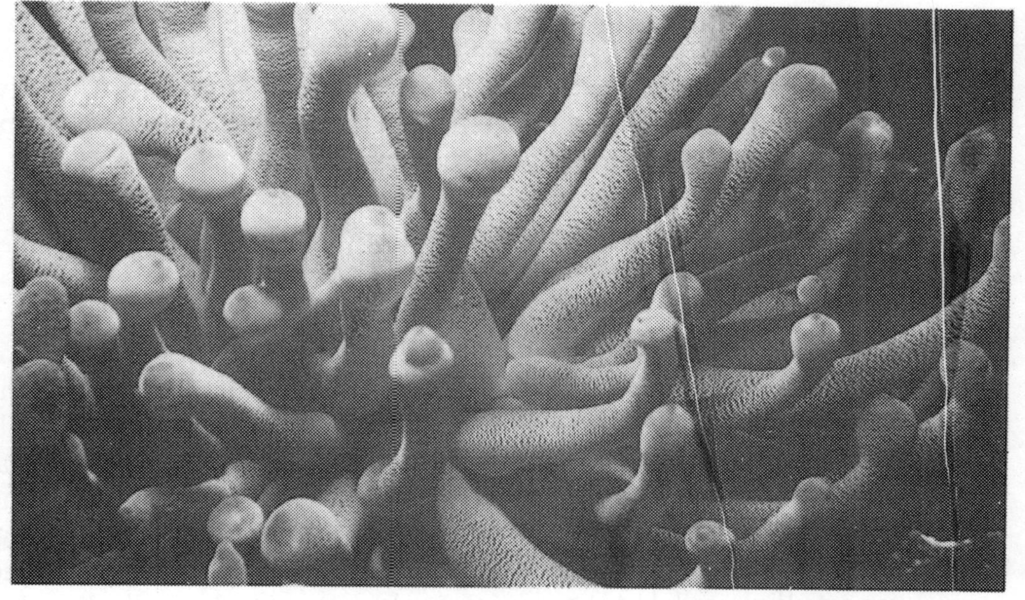

The sea anemone is a relative of the jellyfish, but unlike its cousin, it spends its entire life attached to rocks. This one was found on the ocean floor, off the Cozumel (Mexico) coast. *AP Photo/University of Wisconsin—Superior, Margie Mullet*

Are all **sponges passive**?

No, while most sponges have cells that move about, grabbing and digesting bacteria or debris that float by in the ocean currents, some sponges don't just sit around waiting for microscopic food particles to float by. In the shallow, underwater caves of the Mediterranean Sea, scientists have discovered a new—carnivorous species—of sponge with tentacles, covered with tiny velcro-like hooks. These tentacles snag prey swimming by, usually shrimp-like crustaceans; more tentacles then surround the victim as the sponge begins to digest its catch.

Are the **sea anemone** and **jellyfish related**?

Yes, the sea anemone and jellyfish are very closely related. A sea anemone resembles an upside-down jellyfish, but instead of swimming freely, like its cousin, it spends its entire life attached to rocks.

How much of a **jellyfish is water**?

Even the firmest jellyfish is mostly water—about 94 percent. The jellyfish catches its food using long tentacles that hang from its body; many of these tentacles are armed with stinging cells.

What is a **Portuguese man-of-war**?

One of the most well-known "jellyfish" is the Portuguese man-of-war (*Physalia physalia*), but it is not a true jellyfish, although it is in the same phylum. A man-of-war is actually a colony of similar animals attached to a gas-filled sac at the ocean surface, trailing a mass of tentacles that average 50 feet (15 meters), with the longest recorded being 160 feet (50 meters). The tentacles can give a powerful, painful sting, and have been known to cause a fatal shock to humans. The gas-filled sac acts like a sail, but the man-of-war has no control over its direction of travel—thus you cannot scare the organism away. The overall organism has specific places for reproduction, capturing prey, and digestion. Its predators include sea turtles and janthina snails.

Does the **jellyfish** really have the **longest known reach of any animal**?

Yes, in particular, the Arctic jellyfish has the longest known reach of any animal. These creatures can grow to immense sizes. One picked up on a Massachusetts beach in 1865 measured 7.5 feet (2.3 meters), with 120-foot- (37-meter-) tentacles—giving it a total span, from tentacle to tentacle, of more than 240 feet (74 meters).

How do **jellyfish swim**?

Most jellyfish are shaped like a mushroom, but the weak body wall does not really produce powerful contractions. The animal's rounded body (sometimes called a bell) forces out jets of water diagonally and downward—keeping the animal up and propelling it mostly in a vertical direction.

Is a **comb jelly** a jellyfish?

Bryozoans often live in colonies, such as these that formed around a submerged pipe in Pokai Bay, Oahu, Hawaii; a goatfish is among the visitors. *NOAA*

No, comb jellies are not jellyfish, but they are often mistaken for the other animal. Both comb jellies and jellyfish are gelatinous (thus the word "jelly"), and have a large "head" and tentacles. But there are important differences: Most comb jellies cannot sting, whereas most jellyfish can. Comb jellies are divided into eight sections by bands of hair-like cilia that allow them to move forward and backward through the water; jellyfish move vertically by pulsations of their bell, but they can only move horizontally by the movement of waves and currents.

What is a **bryozoan**?

Bryozoans are mainly ocean invertebrates that gather in colonies and are abundant along coasts, where they might be seen on pilings, shells, rocks, and algae. They are often mistaken for seaweed, moss—or water-logged spaghetti! Most of these organisms measure less than 1/32 of an inch (less than 4/5 of a centimeter) long—their main bodies attached to oval-shaped, box-shaped, or tubular calcareous shells with muscles. The organisms start out as free-swimming larvae that eventually land on a solid structure or rock. A bryozoan's tentacles move in and out of its shell, capturing minute food particles; the organism excretes wastes from an anus tube right next to the tentacles.

What are **brachiopods**?

Brachiopods are similar to bivalve mollusks—they are ocean invertebrates with bivalve shells (in other words, they are soft-bodied but are enclosed within two shells). They are divided into those that have hinged shells and those that have non-hinged shells. The brachiopod has a pair of "arms" inside its shell; these are attached with tentacles, which the animal uses to grab microscopic food particles from the passing water. They were much more prolific in the past; today, they include the lampshells.

What **animals** represent the **Mollusca phylum**?

Although this phylum is under revision (there are disagreements as to divisions within the group), most of the members have familiar names: snails, oysters, clams, mussels, squids, octopuses, and nautiluses. These animals are highly-developed, unsegmented invertebrates. With more than 80,000 different species, the mollusks are a diverse group.

What are **bivalves**, **univalves**, and **cephalopods**?

These terms all describe members of the Mollusca phylum: The bivalves (which are protected by two similar shells, or valves) include clams, mussels, scallops, oysters, and cockles; the univalves (having one shell, or valve, and also called gastropods) include snails, conches, and abalones; and the cephalopods include nautiluses, squids, octopuses, and cuttlefish. Since most people probably equate mollusks with shellfish, this last group requires some explanation: The cephalopod has a group of muscular arms around the front of the head and these arms usually are equipped with series of suckers; these creatures also have highly developed eyes and they usually have bags of inky fluid, which can be squirted as a defense mechanism.

Were **bivalves** really used to make **gloves**?

Yes, as strange as it sounds, these ocean creatures once supplied Italian glove-makers with a thread-like material. The pen shell bivalve, the largest in the Mediterranean Sea, attaches its shell to the seafloor using thick, strong thread. Fishermen would harvest the shell—and the strong thread would be turned over to the leather craftsmen to be woven into gloves.

What are some **characteristics** of **clams**?

Clams are one of the more peculiar species of animals. Because they lack a recognizable head, no one can tell the difference between the front and the back of this bivalve mollusk. They are able to move freely, but they rarely change their locality. The clam secretes enough calcium carbonate to create the two plates—one a bit larger than the other—that surround its main body. Along the hinge line between the two shells, there are usually internal locking teeth. When the shell is closed, there is a slight gap, allowing the passage of its one foot.

How can you tell the **age of a clam**?

Like rings telling the age of the tree, the clam shell has rough striations on its shells, each division representing yearly growth. Some of the rings differ in size depending on the surrounding environmental conditions or seasons—again, similar to trees.

What is a **mussel**?

Mussels are bivalve mollusks—they have two elongated shells, live in intertidal and shallow subtidal zones, and are often found in large groups called mussel clumps. They attach themselves to rock using threads secreted as a liquid from a gland near the foot; the thread hardens on contact with water. They are extremely adaptable and have a worldwide distribution.

The principle predator of the mussel is the starfish; flounder and seabirds also consume mussels. Humans harvest the mussel as a food source, as the animals have a high concentration of vitamins, protein, and minerals.

How does a **scallop move** across the ocean bottom?

Unlike most bivalves, scallops move across the ocean floor very quickly. They don't use a foot, but rapidly open and close their hinged shells, squirting out water that "hops" them across the seabed. They also can

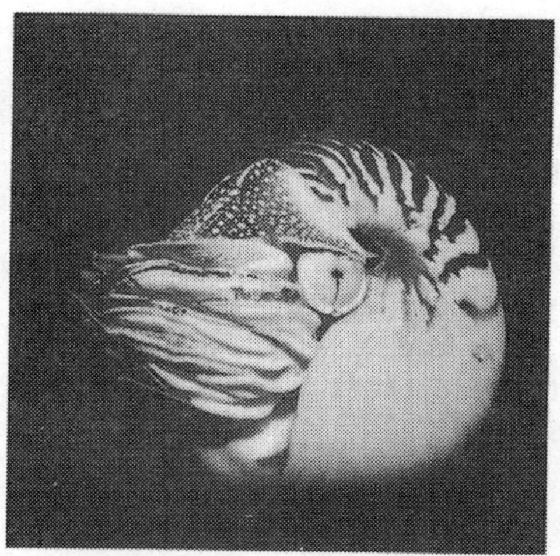

As the nautilus grows, it creates new chambers to produce the spiraling shell for which this primitive-looking mollusk is known. *CORBIS/Amos Nachoum*

see where they are going as they clap their valves—they use the eyes around the edge of their shells.

What are **gastropods**?

The gastropods (univalves or one-shelled) are the largest class of mollusks—about 15,000 fossil forms have been identified, and there are well over 35,000 species living today. Most gastropod shells are corkscrew-shaped; the spirals are called whorls. The tip of the shell, or the apex, is the smallest whorl, laid down in the animal's early life. As the gastropod grows, it lays down more whorls, with the final spiral called the body whorl, ending at the aperture or opening. The gastropods include the tulip snail, abalone, conch, and cowry.

What are the "**butterflies of the sea**"?

Shallow water, shell-less gastropods (univalves) are also known as sea slugs or "the butterflies of the sea." Instead of being protected by shells, these animals have developed biological or chemical methods of defense against predators. But it is their beautiful and vividly colorful warning patterns that give them their nickname. These animals include the sea hare and the nudibranch.

What is a **nautilus**?

The nautilus is a cephalopod (a member of the class Cephalopoda) of mollusks; because of its shell, the chambered nautilus is considered one of the most primitive known living cephalopods. The shell is sectioned into chambers by septa, all made of the same material. As the animal grows, it creates new, larger chambers—thus the increasing spiral of the shell. The nautilus has more than 90 tentacles topped with a hood. It

The squid (seen here in a close-up of the tail end) is the fastest-moving invertebrate: In a form of jet propulsion, it squeezes water out of its body to propel itself through the water. *NOAA/OAR National Undersea Research Program*

also has well-developed eyes, but they are less complex than those of other cephalopods.

There are only 6 known species of nautilus, most living in deep tropical waters—thus scientists know very little about the creatures. What is known is that by day, this animal remains on the ocean floor, either resting or holding onto the bottom with its tentacles. When it swims, it propels itself by forcing water out of its front cavity (similar to all cephalopods) moving the nautilus backward. At night, the nautilus rises into shallower water: By secreting gas into the shell chambers and removing the water, the nautilus floats upward. Once in shallower water, it feeds on reefs or in rocky areas, using its tentacles to capture small, slow moving fish or invertebrates.

What are some **common characteristics of giant squids**?

The giant squid (*Architeuthis dux*) is a huge cephalopod, with 8 arms and 2 long tentacles that end in sucker-lined "clubs," which it uses to grasp prey. No one has ever seen a giant squid alive. The largest to wash up on shore measured 59.5 feet (18 meters) in length and weighed half a ton. They apparently range in size from about 20 to 43 feet (6 to 13 meters) long and weigh 100 to 660 pounds (45 to 300 kilograms). The giant squid is thought to have the largest eyeball of any animal: It measures almost the size of an adult human head! These creatures are hunted by sperm whales; the squid, in turn, eats mainly fish and other squids.

Do **squid** really shoot **ink** to protect themselves?

Actually, both the squid and octopus produce a dark brown ink known as sepia. They squirt the liquid not only when they're scared—but also to confuse predators.

How does a **squid swim** through the open ocean?

A squid—the fastest swimmer among the invertebrates—propels itself by forcefully squeezing seawater out of its body in one direction. This kind of "jet propulsion" allows the animal to quickly catch its primary prey—fish—or even propel itself out of the water, occasionally landing on the decks of ships! Squids have 10 extensions (8 arms and 2 tentacles) with suckers attached, that surround the head; they also have a streamlined body that helps minimize drag through the ocean waters. Even with all this speed and agility, it is thought that a squid must keep swimming or it will sink.

How **fast** can large **squids** swim?

Larger squids, especially over shorter distances, can reach speeds up to 20 miles (32 kilometers) per hour. They are among the fastest of all marine organisms.

Does an **octopus** like to **swim**?

No, considering the octopus is an ocean animal, it doesn't do much swimming. Only when threatened does the octopus swim about. This is because the cephalopod lacks the streamlining that makes other ocean-dwellers good swimmers. The octopus prefers to remain in contact with solid structures, pulling itself along with its sucker-lined arms. An octopus is also a solitary animal, seeking shelter in a cave, den, or under rocks—and because it does not have a shell, it can squeeze into some very minute holes and cracks. These animals usually only come out of hiding to find food or ward off predators.

Most octopus species have about 240 suckers on each of its eight arms—or nearly 2,000 suckers per octopus! *NOAA/OAR National Undersea Research Program; T. Schaff*

How can the **octopus' arm** be described?

The arm of most octopus species have about 240 suckers—usually in double rows. The suckers vary in size from fractions of an inch to 2.8 inches (a few millimeters to 7 centimeters) in diameter. To show how well the suction works, it would take about 6 ounces (170 grams) of pressure to break the hold of a sucker almost 1 inch (2.5 centimeters) in diameter. Imagine what it would take to break the hold of the average octopus! (Assuming 240 suckers at 1 inch, or 2.5 centimeters, each, it would take about 90 pounds, or 41 kilograms, of pressure to break free of this animal's grip.)

Do any **octopuses** have the ability to **produce light**?

Yes, scientists have recently discovered a bizarre deep-sea octopus with suckers that glow in the dark—the *Stauroteuthis syrtensis.* Researchers believe that in the past, the octopus used its suckers for gripping, much like octopuses in shallow water use suckers to grip prey and rocks as they scamper across the ocean floor.

But this "new" bright orange octopus with webbed arms is different. It lives in open water at depths of about 325 feet (900 meters). And because

Horseshoe crabs, like the pair mating here, are the descendants of crustaceans that evolved millions of years ago in the oceans; crustaceans are among the most highly successful groups of animals that have ever lived. *NOAA/Mary Hollinger, NODC Biologist*

it has nothing to cling to, and feeds on prey too small to grab with its arms, it has developed some other tricks. The small, rounded pads on the inside of the octopus' arms look like suckers, but they lack the muscles needed for a good grasp. Instead, the suckers have photocytes, or light-producing cells, that emit a blue-green light. And because most octopuses have good eyesight and use visual signals to communicate, this light may be a good way to attract prey, threaten enemies, or contact others of their species in the darker environment. Scientists call this an evolutionary transition—an evolutionary change that is currently occurring in a species, usually to fill a niche (specific habitat). The only other known example of such a change from muscle to a light organ was in a fish.

Where are **crustaceans classified** in the animal kingdom?

These familiar creatures, including the crab and lobster, are members of the class Crustacea in the phylum (group) Arthropoda. In fact, since there are more than 26,000 species of crustaceans they are the largest class of arthropods. The crustaceans, most of which are marine, have chitinous (fibrous) and/or calcareous (shell-like) exoskeletons; in other words, they wear their somewhat soft skeletal structures on the outside of their bodies. These animals have segmented bodies, usually with a pair of legs (or appendages) per segment. They also have two sets of antennae. Other marine crustaceans include the sea spiders, krill, and barnacles. (Their land relatives include wood lice.)

What is **chitin**?

Chitin is the key component (a tough polysaccharide) of the exoskeletons of arthropods and the shells (also really exoskeletons) of crabs, lobsters, and insects. It is also present in the hard structures of some coelenterates, hard corals, and sea anemones. Chitin is actually a natural polymer that is structurally related to the sugar called glucose; it has been made into water-resistant paper and edible food wrapping. Scientists have even developed a technique to manufacture string-like strands of chitin—for use as dressings for cuts and burns, as the chitin apparently has antifungal properties and promotes healing.

Using a gland on its head, the barnacle will "glue" itself to any surface, spending its life there in an upside-down position. Here, gooseneck barnacles have attached themselves to a dead sponge. *NOAA/OAR National Undersea Research Program; J, Moore*

What are **sea spiders**?

Sea spiders are somewhat similar to spiders on land, but these 4- to 6-legged arthropods—also called whip scorpions—live in the oceans. The smaller specimens measure about 0.01 inches (2 to 3 millimeters) and live in shallow waters; the larger ones, some measuring more than 20 inches (50 centimeters) in length, live in deep-ocean waters. They are carnivorous, some feeding on other invertebrates, sucking them dry, or tearing apart prey with their legs.

What **marine animal** spends its life **standing on its head**?

The crustaceans called barnacles spend their entire adult lives on their heads: The barnacle larva is free swimming, but when it finds a suitable spot to live, it glues itself to the surface using a gland on its head. The barnacle then builds shell plates around its upside-down body in a volcano-like shape—thus as an adult, the animal becomes immobile.

Many barnacle species live in intertidal zone, where they are alternately covered and uncovered by the tides. When the tide is out, the barnacle plates are shut tight, holding in water; when the tide comes in, the plates open at the top, and a feathery "hand" (remnants of the arthropod's leg) grabs the passing food particles, then passes the food into its stomach.

Where do **barnacles attach** themselves?

Not all barnacles attach themselves to rock, as many boat owners can verify. Some species of barnacles are parasites of other crustaceans or of corals; others are independent species that live in communities attached to whales, turtles, and other marine organisms. Others have a negative impact on commercial marine enterprises, covering the bottoms of ships, piers, and offshore installations.

What are the **"edible" crustaceans**?

The edible crustaceans are names familiar to most of us, such as krill, crab, lobster, shrimp, crawfish, and crayfish. But these animals really only represent about 3 percent of the total worldwide marine catch (fish total about 90 percent of the catch).

Ancestors of today's crustaceans evolved millions of years ago in the oceans, and are considered some of the most highly successful groups of animals that have ever lived. Of these species, only a relatively few are used as food sources for humans, food or bait for other marine animals, or fertilizer for crops.

What are **krill**?

Shrimp-like krill are the largest plankton in the rich upper layer of the ocean, and some of the smallest crustaceans. Most krill are found in the South Atlantic Ocean off Antarctica; in the Norwegian Sea; and in other cold ocean waters around the world.

The krill differs from the shrimp in one important way: It has bristles at the end of its tail. Krill are often found in huge groups, numbering almost 100 million individuals in one region. Krill have an important role at the bottom of the marine food chain: They are food for certain fish, seals, penguins, seabirds, herring, and sardines; giant blue and baleen whales sometimes feed exclusively on krill. In fact, it is estimated that baleen whales consume about 33 million tons annually; penguins, about 39 million tons; and fur seals, about 4 million tons.

Krill is also harvested by a number of countries. Japan has been known to produce krill meat and protein concentrate. Russia has also been involved in krill catches, making such products as krill butter and cheese.

Can **krill populations** indicate a **climate change**?

Yes, krill populations may indicate climate changes—and other environmental changes as well. The tiny animals are extremely sensitive to fluctuations in temperature, salinity, and ultraviolet radiation. Scientists have been watching the changes in the krill populations, especially as any reductions correlate to recorded decreases in the Southern Hemisphere's ozone layer—the layer that protects the Earth's organisms from the ultraviolet rays of the Sun.

Where are the **major crab fisheries** around the world?

Major crab fisheries are found along the Asian and North American coasts bordering the central and northern Pacific Ocean (where king and Dungeness crabs dwell) and along the Atlantic coast of North America. The Bay of Biscay, off the northern coast of Spain and the western coast of France; the North Atlantic, off the southern coast of Ireland; and the North Sea are also prime spots for crab fishing.

The oldest crab industry in the United States began with the blue crab (*Callinectes sapidus*), with records mentioning the crab in the Chesapeake Bay as early as the 1630s. Today, the blue crab is often found off Florida, Maryland, North Carolina, and Virginia.

What are **lobsters** and where are they **found**?

Lobsters are crustaceans, and include the true, spiny, Spanish (or slipper) and deep-sea lobsters. They are part of the order Decapoda, which, in addition to lobsters, includes the other crustaceans that are most often eaten by humans—crabs and shrimp. All lobsters have

Lobsters are most abundant in about 1,000 feet (300 meters) or more of water; this rock lobster was found on a Pacific reef. ***NOAA/OAR National Undersea Research Program; E. Williams***

stalked eyes; chitinous (fibrous), segmented exoskeletons (on a three-part body); and five pairs of walking legs. One pair of walking legs is the chela, or claws—and in most cases, one claw is larger than the other.

Lobsters are very long-lived animals, maturing at about age 5 and living more than 50 years. They usually survive that long because they have few predators; if they can stay away from starfish or rays, their only predator is man. Lobsters are most abundant in about 1,000 feet (300 meters) or more of water; they live on carrion (dead animals), or even live fish when they can catch them.

What is known about **lobster behavior**?

Lobsters seem to be their own animal: No one has been successful "growing" and harvesting lobsters commercially, which is why lobster cages are still found all along the coasts of the world. The behavior of this crustacean often seems bizarre, too. For example, North Atlantic lobsters have been seen traveling in columns along the ocean floor—but no one knows why. Some scientists have theorized that the marches are somehow associated with the animal's rise in abundance.

Which **lobsters** are **harvested**?

There are three species of lobsters that are harvested the most: the American lobster (*Homarus americanus*), representing about 50 percent of the catch; the European lobster (*H. gammarus*), about 30 percent; and the Norwegian lobster (*H. norvegicus*) making up most of the rest.

What is a **shrimp**?

Shrimp, of which there are more than 2,000 known species, are among the smaller crustaceans (hence the association of the word with anything that is small). They inhabit environmental niches (habitats) from freshwater rivers to the deep oceans, but most shrimp are ocean-dwellers. The body is somewhat flat, and has a tail fan attached to a rather long abdomen. Most shrimp spend their time walking around the bottom of the ocean on their 5 pairs of walking legs, or they use the limbs to dig up sediment to loosen food particles. They usually feed at night on other small crustaceans, worms, and mollusks.

Most shrimp live an average of 3 years. But they make up for a short lifetime with their prolific reproduction: In 3 years, a female shrimp can produce more than 20,000 offspring. Shrimp are also more resilient than are other crustaceans—they can tolerate drastic changes in the temperature and salinity of local waters.

Why are **shrimp** called **"bandits of the coral reefs"**?

Not all shrimp are bandits; this nickname refers specifically to the 2-inch (6-centimeter) pistol shrimp, which dwells in and around coral reefs. It

Do sea urchins throw their spines?

No, it's a myth that sea urchins throw their spines. They also do not leap onto helpless people passing by. And most sea urchins can be picked up and held, if it's done carefully. But there are exceptions. For example, the long-spine sea urchin of south Florida and the Caribbean has spines that can easily penetrate human skin, breaking off (like a splinter). And because the spines' barbs are slightly toxic, they can be quite painful.

Sea urchins are often found in tidal pools or just below the low-tide line—thus they are prey of seabirds, sea stars, lobsters, and terrestrial animals such as foxes. The globe-shaped body of a sea urchin is called a test; the mouth is on the bottom and the anus on top. The test is divided into 10 sections, 5 of which have holes through which tube feet protrude (in rows). A sea urchin eats using a structure in the mouth called Aristotle's lantern; at the center are 5 teeth that come together like a bird's beak. These teeth allow the animal to scrape algae off rocks; as the teeth wear down, they grow back (maintaining their size). The spines that cover the body are used for cleaning and defense, and some contain poison. Each spine is connected to the test with a ball-and-socket arrangement, allowing the spines to move. When a sea urchin dies, the spines drop off and the body inside the test decomposes, leaving only the shell.

sports a large right pincher claw that has a peg and matching hole. When a small fish wanders by, the tiny crustacean runs out of hiding, aims its "pistol," and snaps its large pincer shut. This creates a shock wave, stunning the fish—and allowing the shrimp to move in for the kill.

Do some **fish** really **protect shrimp**?

Yes, a certain type of goby fish, the *Cryptocentrus coeruleopunctatus,* stands guard over a snapping shrimp at the entrance to their shared bur-

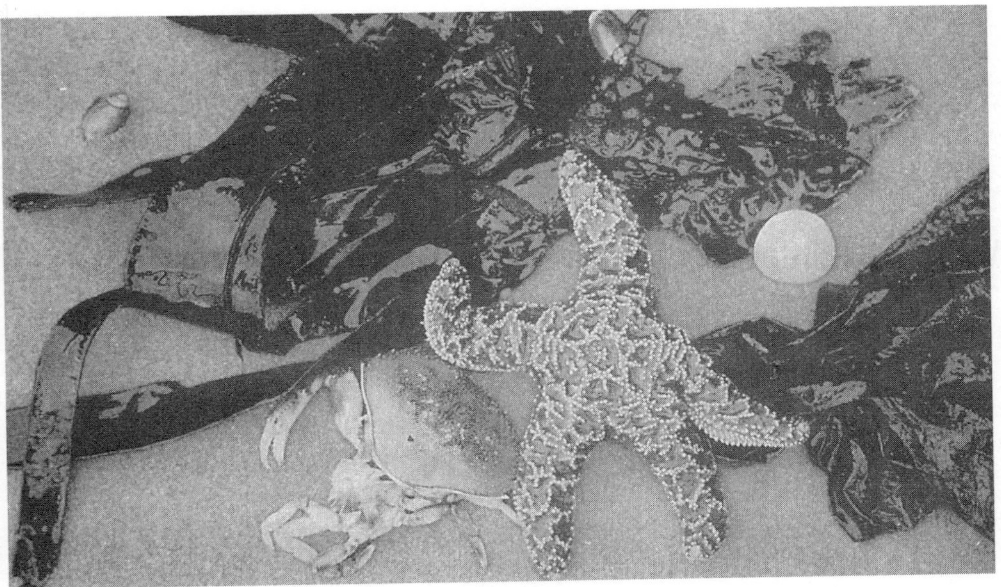

At low tide in central Oregon, a sampling of sea life gets stranded on the beach: kelp, a crab, a sea star, and a sand dollar. *CORBIS/Brandon D. Cole*

row on the ocean floor. As the tiny shrimp digs and cleans the burrow with its claws, the goby stands guard, one of its antennae touching the shrimp. If there is danger, the goby wiggles. The shrimp, feeling the movement in the antenna, runs for cover—with the goby not far behind.

What is the difference between **crayfish** and **crawfish**?

Crayfish are crustaceans that usually live in freshwater; crawfish live almost exclusively in the oceans (or in brackish waters). Because crawfish resemble their crustacean cousin, the lobster, they are sometimes called "spiny lobsters" or "false lobsters"; but crawfish are not really lobsters, though many crawfish are just as large—and certain species are considered a delicacy.

What is a **sand dollar**?

A sand dollar is actually a flat version of a sea urchin. It has very tiny moveable spines on the test (body) that give the sand dollar a smooth, felt-like appearance and touch. These spines allow the animal to dig into the sand. The flower-like shape on the top of the sand dollar corresponds to the sea urchin's 5 rows of tube feet. The tube feet that pop out of these petals are used for respiration (breathing).

What is the **difference** between a **starfish** and a **sea star**?

There really is no difference: Starfish is an antiquated and inaccurate name that is now being replaced with sea star (since a starfish is not a fish at all). Sea stars, members of the phylum (group) Echinodermata, most often live in rock tidal pools, and measure 6 to 12 inches (15 to 30 centimeters) in diameter, although some have been found up to 26 inches (65 centimeters) in diameter. Most sea stars have 5 arms, although some have up to 10. The top of the sea star consists of horny (chitinous) plates; the undersides of the arms have tubed feet, or a series of small, flexible suction cups. It uses its arms to move and grab prey; it can also regenerate its arms. The mouth of this predator is found at the center of the star, on the bottom. Worms, crustaceans, and bivalves (especially oysters) are among the foods consumed by the sea star.